W9-AAE-206

Progress in Mathematics
Volume 153

Jorge Buescu

Exotic Attractors

From Liapunov Stability to Riddled Basins

Birkhäuser Verlag
Basel · Boston · Berlin

Author:

Jorge Buescu
Instituto Superior Técnico
Departamento de Matemática
Av. Rovisco Pais
1096 Lisboa Cedex
Portugal

1991 Mathematics Subject Classification 58F13, 58F11

A CIP catalogue record for this book is available from the Library of Congress, Washington D.C., USA

Deutsche Bibliothek Cataloging-in-Publication Data

Buescu, Jorge:
Exotic attractors : from Liapunov stability to riddled basins / Jorge Buescu. – Basel ; Boston ; Berlin : Birkhäuser, 1997
(Progress in mathematics ; Vol. 153)
Zugl.: Warwick, Univ., Diss.
ISBN 3-7643-5793-2 (Basel ...)
ISBN 0-8176-5793-2 (Boston)

Library of Congress Cataloging-in-Publicataion Data

Buescu, Jorge. 1964–
Exotic attractors : from Liapunov stability to riddled basins /
Jorge Buescu.
p. cm. -- (Progress in mathematics ; v. 153)
Includes bibliographical references and index.
ISBN 3-7643-5793-2 (alk. paper). -- ISBN 0-8176-5793-2 (alk paper)
1. Attractors (Mathematics) I. Title. II. Series: Progress in mathematics (Boston, Mass.) ; vol. 153.
QA814.813.B84 1997
514'.74 -- dc21

Printed on acid-free paper produced of chlorine-free pulp. TCF ∞
Printed in Germany
ISBN 3-7643-5793-2
ISBN 0-8176-5793-2

9 8 7 6 5 4 3 2 1

The case for my life, then, or for that of anyone else who has been a mathematician in the same sense in which I have been one, is this: that I have added something to knowledge, and helped others to add more; and that these somethings have a value which differs in degree only, and not in kind, from that of the creations of the great mathematicians, or of any of the other artists, great or small, who have left some kind of memorial behind them.

G. H. Hardy, *A Mathematician's Apology.*

To both of you.

Contents

Preface

This book grew out of the work developed at the University of Warwick, under the supervision of Ian Stewart, which formed the core of my Ph.D. Thesis. Most of the results described were obtained in joint work with Ian; as usual under these circumstances, many have been published in research journals over the last two years. Part of Chapter 3 was also joint work with Peter Ashwin. I would like to stress that these were true collaborations. We worked together at all stages; it is meaningless to try to identify which idea originated from whom.

While preparing this book, however, I felt that a mere description of the results would not be fitting. First of all, a book is aimed at a wider audience than papers in research journals. More importantly, the work should assume as little as possible, and it should be brought to a form which is pleasurable, not painful, to read.

These are the reasons why I decided to include historical remarks at the end of each chapter. In order to make the book self-contained, a thorough review of the essential concepts is needed from the very beginning. Why not complement the presentation of the concepts as we know them today with an overview of the way they originated and evolved?

A cautionary note, though. The above comments should not mislead the reader into thinking that this book is intended to be a systematic development or a thorough review of specific areas of Dynamics. Let me be clear about this: it is *not*. It is a research report on two different problems in Dynamics, the largest common divisor of which is the concept of attractor of a dynamical system. This fact emerges at the most apparent level of the book: its title. Thus "exotic attractors" are not a class of attractors with specific properties; the adjective "exotic" is not a mathematical characterization. It merely indicates that in both problems the attractors in question, not necessarily related to each other, have certain distinctive features which may or may not be a part of the intuitive mental picture of an attractor.

Every time some mathematical structure is added, care is taken to explore its consequences. However, concepts and their relationships are only developed up to the point of their usefulness to the problems at hand. This should not be taken to mean that further developments do not exist or are not useful.

I would like the book to reflect its genesis in research, describing the positive results and living topics but also the dead ends and open problems. I decided, besides reporting where we stand, not to delete the footprints left on the way. It is bad enough that Mathematics often presents to outsiders the image of a polished, perfect, finished body of absolute truths. We all know Mathematics isn't so: I didn't pretend it is when writing this book. Sometimes conciseness pays the price of this option, as the reader will undoubtfully appreciate in Chapter 2.

I don't know if these goals have been achieved in a balanced and articulate manner; only the reader can be the judge of that. I can only hope that he enjoys reading this work as much as I enjoyed writing it.

Chapter 1 consists of a thorough review of the concepts used in later chapters, assuming very little at the outset – in fact nothing beyond the definition of a (semi-)dynamical system. Several results are established on the nature of Liapunov and asymptotic stability and on the implications of transitivity (i.e. existence of a dense orbit) for the topology of the connected components of an invariant set. Some relations between the topological and ergodic properties of invariant sets and attractors are described. We then shortly discuss Axiom A, asymptotically stable and Milnor attractors. Counterexamples are given to illustrate the finer points of the results and definitions.

In Chapter 2 the first body of results is described. Suppose A is a Liapunov-stable transitive set with infinitely many connected components – a prototype of which is the Cantor set arising at the Feigenbaum limit of period-doubling. By the results in Chapter 1, the quotient by components generates a Cantor set K on which the induced map is transitive. The main Theorem states that this map is an adding machine. This result is shown to hold under the weaker assumption that A is a stable ω-limit set by the more elegant inverse limit constructions. These results, which in a sense are surprising since they lie close to the foundations of Dynamics, have non-trivial consequences. For interval maps they imply that every point in the adding machine is a limit point of periodic points whose periods are very naturally determined by the adding machine itself.

In Chapter 3 the following problem is addressed. Suppose a dynamical system possesses an invariant submanifold restricted to which it has an asymptotically stable attractor A. When is A an attractor for the full system, and in what sense? This question is particularly meaningful when A is chaotic, and has arisen under different guises (e.g. synchronization problems or systems with symmetries) in applications-oriented literature over the last decade. The dynamics near A for the full system is characterized by the spectrum of normal Liapunov exponents. This spectrum determines the points where A: (1) ceases to be asymptotically stable, possibly developing a riddled basin; (2) ceases to be an attractor; (3) becomes a transversely repelling chaotic saddle. A sufficient condition for the creation of a riddled basin in transition (1) is provided. With these results an adequate bifurcation theory is developed. We study in some detail analytical and physical systems exhibiting this behaviour, including an experimental system of symmetrically coupled electronic oscillators.

My greatest debt is of course to Ian Stewart. Ian didn't offer me a fish: he taught me how to do my own fishing. He was always there when I needed him but also gave me the freedom to pursue my own ideas – and make my own mistakes. Needless to say, if any have survived up to this stage (which I hope not) only I am to blame.

The mathematical atmosphere at Warwick was always stimulating. I have benefited enormously from regular discussions with many Warwick mathematicians, notably Greg King, Robert MacKay, Mario Micallef, David Mond, Mark Pollicott, David Rand and especially Peter Walters. I also had important discussions for the brewing of these ideas with Ian Melbourne. It was a great pleasure to work with Peter Ashwin; many fine points only became clear after challenging but fruitful brainstormings.

Many people helped me in many ways during these years. My Mother gave me constant and unfailing support. I have met wonderful people at Warwick, without whom my work would have been much less bearable. Their friendship often brought brightness to the darker hours; I thank them all now that my work has seen daylight. I would like to thank especially Inês Cruz and Carlos Monteiro for their friendship, Claudio Arezzo for the long discussions of Mathematics and Pat and Lito de Baubeta for everything.

I also thank the publishers at Birkhäuser for showing a keen interest in my work and for their encouragement – which includes their patience at the final stage of preparation of the book!

The research leading to this work was conducted while on leave from the Departamento de Matemática of Instituto Superior Técnico. During this period I had partial support from JNICT-CIENCIA through grant BD/1073/90-RM and from Fundação Calouste Gulbenkian.

The work leading to this book was partially supported by projects JNICT-PBIC/P/MAT/2140/95 and JNICT–PRAXIS/2/2.1/MAT/199/94.

My years at Warwick also witnessed the happiest moments in my life: the blossoming of a family. The time has come to thank Catarina and Henrique for all their sacrifices.

List of Figures

Chapter 1
Attractors in Dynamical Systems

1.1 Introduction

The notion of attractor in a dynamical system is an underlying theme throughout this work. In a sense, it is the unifying concept which links the various topics in dynamics which are dealt with in this book.

Although the concept of attractor is often taken for granted, the fact is that several distinct, sometimes overlapping, definitions of attractor exist in the literature. Moreover, each of these definitions illuminates specific properties of the dynamical objects under consideration. We interpret this state of affairs as follows. There is no contradiction between the several distinct concepts of attractor – all of them are useful as they reveal slightly different properties. The characterization of the object is given by the set of all these properties, not just by a specific one. We therefore think there is no such thing as a "final" or "correct" definition of attractor.

However, in mathematics it is essential to work with rigorously defined terminology. We therefore start this book with the precise definition of the main concepts we use, establishing their properties in parallel.

The organization of this chapter is as follows. We start by outlining the general framework – namely the topological and differential structures – in which all the work is developed, and by giving the basic mathematical definitions, mostly from topological dynamics, from first principles. In § 1.3 we state and prove general results which focus on different aspects of these concepts. These highlight some properties of Liapunov stability, asymptotic stability, the topology of components of transitive sets, and transitive compact sets with non-empty interior. In § 1.4 we discuss and distinguish between several different notions of attractor which will be relevant for the main work; even though other definitions are possible (and some of those are briefly described in § 1.6) we limit the discussion to the ones which are used at some stage in this book. The following § 1.5 is a pot-pourri of examples and counterexamples relating to the concepts and properties introduced in the previous sections. The chapter ends with a final section presenting general comments and historical remarks.

1.2 Basic definitions

Throughout this book we always assume that (X, d) is a complete metric space and $f : X \to X$ is a continuous map (with respect to d). The pair (X, f) then defines a topological dynamical system to which all the concepts in this section apply.

At this point we introduce two cautionary notes. First, the concepts presented below make sense in more general contexts such as topological spaces, sometimes with some extra uniform structure. However, the weakest structure required by the main results in this work is that of complete metric space. Thus for our purposes the gain in generality in working with the more abstract structure is non-existent and only confuses the real issues. Secondly, we will very often work in the category of differentiable dynamics; in that case X will be a differentiable manifold and f a differentiable map. For several purposes it is extremely useful to endow X with a Riemannian metric ρ; this is essential, for instance, in Chapter 3. By a standard theorem of differential geometry (see Spivak [115], I, Theorem 9.7) the Riemannian manifold (X, ρ) is homeomorphic to X with its original topology – so that there is no real gain or loss of generality when speaking of the properties of f from the topological dynamics point of view through its Riemannian structure.

With this proviso, we now proceed to present the basic concepts of topological dynamics which will be used throughout this work. Most of these are standard from the literature.

Definition 1.2.1 (forward invariance) *A set $A \subseteq X$ is said to be* forward invariant *if $f(A) \subseteq A$.*

If A is forward invariant, then trivially $f^n(A) \subseteq A$ for all $n \geq 0$.

Definition 1.2.2 (invariance) *A set $A \subseteq X$ is* invariant *if $f(A) = A$.*

Obviously invariance implies forward invariance but not conversely. If A is invariant, then $f^n(A) = A$ for all $n > 0$. We also remark that $\bigcap_{n\in\mathbb{Z}} f^n(A) = A$.

Definition 1.2.3 (ω-limit set) *Let $x \in X$. The ω-limit set $\omega(x)$ is the set of accumulation points of the forward orbit $\{f^n(x)\}_{n\in\mathbb{N}}$.*

Observe that the ω-limit set can clearly be written

$$\omega(x) = \bigcap_{m\geq 0} \overline{\bigcup_{n\geq m} f^n(x)}. \tag{1.1}$$

Proposition 1.2.4 *Given $x \in X$, the ω-limit set $\omega(x)$ is closed and forward-invariant. Moreover, if X itself is compact, then $\omega(x)$ is also non-empty, compact and invariant.*

Proof. See, for instance, Block and Coppel [14]. □

Remark 1.2.5 If X is not necessarily compact but $A \subseteq X$ is a compact forward-invariant set, it follows that for all $x \in A$ $\omega(x)$ is non-empty, compact, invariant and contained in A. □

Definition 1.2.6 (topological transitivity) *A compact invariant set A is said to be* topologically transitive *if there exists $x \in A$ such that $\omega(x) = A$.*

In Definition 1.2.6 invariance of A is not a necessary assumption; if omitted it becomes a consequence of A being an ω-limit set together with Proposition 1.2.4. Thus there is no loss or gain of generality by assuming it at the outset.

We shall refer to topological transitivity as transitivity for short, when no ambiguity is possible.

We will show in the next section that, if A is transitive, almost all points – in the topological (Baire category) sense – of A satisfy $\omega(x) = A$. This leads us naturally to the next definition, which deals with the extreme case where $\omega(x) = A$ for *all* $x \in A$.

Definition 1.2.7 (minimality) *Let A be a compact transitive set. A is said to be* minimal *under f if $\omega(x) = A$ for all $x \in A$.*

Remark 1.2.8 In § 1.4 we shall use the term *minimal* in another sense. Whenever ambiguity may arise from the context, we shall refer to minimality as in Definition 1.2.7 as *topological minimality*. When no confusion is possible, we abbreviate it to minimality. See also Remark 1.4.10 and § 1.6. □

The term *minimal* is appropriate: the condition of minimality implies that, for every $x \in A$, the smallest ω-limit set containing x is the whole of A, or in other words that A contains no compact invariant non-empty proper subsets.

We now turn to topological definitions of dynamical stability.

Definition 1.2.9 (Liapunov stability) *A compact invariant set A is* Liapunov stable *if, for every open neighbourhood U of A, there exists an open neighbourhood $V \subset U$ such that $f^n(V) \subset U$ for all $n \in \mathbb{N}$.*

Definition 1.2.10 (Asymptotic stability) *A compact invariant set A is said to be* asymptotically stable *if it is Liapunov stable and there is an open neighbourhood W of A such that $\omega(x) \subset A$ for all $x \in W$.*

Definition 1.2.11 (Basin of attraction) *Let A be a non-empty compact invariant set. The set*

$$\mathcal{B}(A) = \{x \in X : \ \omega(x) \subset A\}$$

is called the basin of attraction *of A.*

Note that Definition 1.2.11 implies that the basin of attraction is always non-empty, as it contains at least A. On the other hand, this definition is completely unrelated to stability properties of A; the noun 'attraction' in Definition 1.2.11 does *not* mean that A is an 'attracting set' in any sense.

It follows from the previous definitions, however, that asymptotically stable sets have an open basin of attraction: given any W as in Definition 1.2.10, it is easily checked that $\mathcal{B}(A) = \bigcup_{n=0}^{\infty} f^{-n}(W)$ is open and independent of W.

Definition 1.2.12 (Topological conjugacy) *Let (X_1, f_1) and (X_2, f_2) be topological dynamical systems. We say they are* topologically conjugate *if there exists a homeomorphism $h : X_1 \to X_2$ such that the diagram*

$$\begin{array}{ccc} X_1 & \xrightarrow{f_1} & X_1 \\ h\big\downarrow & & \big\downarrow h \\ X_2 & \xrightarrow{f_2} & X_2 \end{array}$$

commutes.

We shall sometimes say that a dynamical system (X_1, f_1) is *semi*-conjugate to (X_2, f_2) if there is a continuous surjection h – no longer a homeomorphism – making the above diagram commute. Thus h may "collapse" several points of X_1 into the same point of X_2. Note that, as opposed to conjugacy, semi-conjugacy is not an equivalence relation.

In the following definition X will be a compact differentiable manifold. We will denote by $C^r(X)$ the space of all differentiable maps $f : X \to X$ of class C^r equipped with the C^r norm. It is a standard fact that $C^r(X)$ is a complete metric space, see e.g. Hirsch [47].

Definition 1.2.13 (Structural stability) *Let X be a compact differentiable manifold and $f \in C^1(X)$. We say that f is* structurally stable *if there is a neighbourhood $V(f)$ in $C^1(X)$ such that, if $g \in V(f)$, then g is topologically conjugate to f.*

This is sometimes called C^1-structural stability because of the use of the C^1 topology. A similar definition using C^r vector fields and the C^r norm may be given; the resulting notion of stability is then called C^r-structural stability. In this work we shall only deal with the C^1 case and follow the standard practice of referring to it simply as structural stability.

We next introduce some essential concepts from ergodic theory.

Definition 1.2.14 (invariant measure, ergodic measure) *An* -invariant measure *μ is a probability measure defined on the Borel σ-algebra of X with the property that $\mu(f^{-1}(B)) = \mu(B)$ for all measurable B. If $f^{-1}(B) = B \pmod 0$ implies $\mu(B) = 0$ or 1, then μ is ergodic.*

Here the notation $A = B \pmod 0$ means that the symmetric difference $A \triangle B$ has zero μ-measure.

Remark 1.2.15 (existence of ergodic measures) It is a standard fact that, if X is a compact metric space and $f : X \to X$ is a continuous map, there exists at least one ergodic measure supported in X; see Mañé [65], Denker, Grillenberger and Sygmund [27], Pollicott [95] or Walters [118]. □

An extreme case of ergodicity is when there exists a *unique* invariant measure:

Definition 1.2.16 (unique ergodicity) *Let X be a compact metric space and $f : X \to X$ be a continuous map. We say that f is* uniquely ergodic *if there is a unique Borel f-invariant probability measure.*

The set of ergodic measures coincides with that of extremal points of the space of invariant measures (equipped with the weak* topology); see Mañé [65]. In the case of unique ergodicity the space of invariant measures has a unique element μ, which is therefore ergodic.

The *support* supp μ of an abstract finite measure μ defined on some σ-algebra of X is the smallest closed subset of X of full measure (see Rudin [99]).

1.3 Topological and dynamical consequences

In this section we collect together several properties of the concepts introduced above. It is stressed that, while all the results are elementary, some of them are original and some are proved by other authors. Some are even dynamics 'folklore'. Proofs are provided when the results are original or not standard in the statement or in the proof.

Lemma 1.3.1 *Let A be a compact invariant set. The following are equivalent:*

1. *A is topologically transitive.*
2. *$\omega(x) = A$ holds for a residual set in A.*
3. *For any open $U \subset A$ the set $\bigcup_{n\geq 0} f^{-n}(U)$ is dense in A.*

Proof. See Mañé [65], Prop. 1.11.4. □

Corollary 1.3.2 *Let A be a compact transitive set. Then every closed invariant proper subset of A is nowhere dense in A.*

Proof. If $B \subset A$ is closed invariant, then $\omega(x) \subset B$ for all $x \in B$. As $B \neq A$, the result follows from (2) in Lemma 1.3.1. □

The following well-known proposition, which we include for completeness, establishes a characterization of the dynamics in the supports of invariant and ergodic measures in topological terms. The compactness assumption is transferred to A, so that X is not necessarily compact.

Proposition 1.3.3 *Let $A \subseteq X$ be a compact invariant set. Let μ be an invariant measure supported in A; denote by $S_\mu \subset A$ its support. Then:*

1. *S_μ is a compact invariant set.*
2. *If in addition μ is ergodic, then S_μ is topologically transitive.*
3. *If $f_{|S_\mu}$ is uniquely ergodic, then S_μ is minimal.*

Proof. (1) From the definition

$$S_\mu = \bigcap_{\substack{F \text{ closed} \\ \mu(F)=1}} F$$

S_μ is closed, hence compact. We first prove $f(S_\mu) \subset S_\mu$. Suppose for a contradiction that $f(S_\mu) \setminus S_\mu \neq \emptyset$. Take $y \in f(S_\mu) \setminus S_\mu$ and a small enough ball $B_r(y)$ such that $S_\mu \cap B_r(y) = \emptyset$. Then $\mu(B_r(y)) = \mu(f^{-1}(B_r(y))) = 0$. But $f^{-1}(B_r(y))$ is open, and therefore $S_\mu \setminus f^{-1}(B_r(y))$ is closed. It is also strictly smaller than S_μ and has unit measure, contradicting the definition of S_μ. To prove the converse inclusion, suppose $f(S_\mu) \subset S_\mu$ strictly. Again $f(S_\mu)$ is compact and strictly smaller than S_μ. Now $\mu(f^{-1}(f(S_\mu))) = \mu(f(S_\mu))$; but $f^{-1}(f(S_\mu)) \cap S_\mu = S_\mu$, hence $\mu(f(S_\mu)) = 1$, contradicting the definition of support.

(2) If $U \subset S_\mu$ is any (relatively) non-empty open set, then $\mu(U) > 0$ – otherwise $S_\mu \setminus U$ would be a closed proper subset of S_μ with full measure, contradicting the definition of support. Let μ be an ergodic invariant measure, and consider an arbitrary non-empty, (relatively) open subset $U \subset S_\mu$. The open subset $W = \bigcup_{n=0}^{\infty} f^{-n}(U)$ is forward-invariant, and by ergodicity it has full measure. Thus $\mu(W^c) = 0$, which implies int $W^c = \emptyset$; therefore W is dense in S_μ, which is equivalent to topological transitivity by Lemma 1.3.1.

(3) Let $\Lambda \subset S_\mu$ be a non-empty compact invariant subset. Then by Remark 1.2.15 there exists an ergodic probability measure ν supported in Λ. Extension of this measure to S_μ (that is, construction of the measure ν_* in the Borel σ-algebra of S_μ satisfying $\nu_*(B) = \nu(B \cap \Lambda)$ for each Borel set B in S_μ) yields an invariant measure ν_* supported in S_μ. By unique ergodicity $\nu = \mu$. Thus supp $\mu = S_\mu$ and therefore $\Lambda = S_\mu$. Therefore S_μ contains no invariant non-empty proper subset; by Proposition 1.2.4 it follows that $\omega(x) = S_\mu$ for all $x \in S_\mu$, proving minimality. □

Remark 1.3.4 The converse to these results is not true; a compact invariant set A is not necessarily the support of an invariant measure, a compact transitive set is not necessarily the support of an ergodic measure and a minimal set is not necessarily the support of a unique ergodic measure (however, in the last case the supports of distinct ergodic measures must obviously be the whole minimal set; see Furstenberg [32] or Mañé [65] for an example using an analytic diffeomorphism of the torus). The reason, loosely speaking, is that statements about ω-limits and transitivity deal with arbitrary subsequences of orbits, while statements about invariant measures and ergodicity deal only with subsequences whose relative frequencies approach a non-zero limit. So, in general, a statement in ergodic theory will be stronger than the (categorically) analogous statement in topological dynamics; if a statement is true in the ergodic context, we should expect the corresponding topological statement to be true – but not the other way around.

It is possible to construct examples where non-empty compact pieces of a transitive set are approached with zero asymptotic relative frequency and thus the corresponding ergodic measure is zero on those pieces. An example which very nicely illustrates this point is derived from a continuous unimodal map f of the interval $[0, 1]$ as constructed in Lasota and Yorke [60]. In this map almost all points, in the Lebesgue measure sense, satisfy $\omega(x) = [0, 1]$ and their Birkhoff averages $\frac{1}{n}\sum_{i=0}^{n-1} \delta_{f^i(x)}$ converge to the Dirac invariant measure δ_0 concentrated on the fixed point 0. □

Our next results require in principle some kind of compactness of the ambient space X. While compactness of X is sufficient, it is not necessary for their validity. We widen the scope of these results by replacing compactness of X by the weaker condition of local compactness – which allows us to encompass, for instance, the cases where X is $\mathbb{R}^n$ or more generally a non-compact finite-dimensional manifold. That this condition will be sufficient for our purposes is the consequence of an elementary proposition in point-set topology which we state and prove below.

Proposition 1.3.5 *Let X be a locally compact metric space and $A \subset X$ be a compact set. Then any open neighbourhood U of A contains an open neighbourhood V of A consisting of a finite number of disjoint non-empty open sets with compact closure.*

Proof. Consider the open cover of A given by $\mathcal{O} = \{B_\delta(x) : x \in A\}$, where $\delta = \delta(x)$ is such that, for all $x \in A$, $B_\delta(x) \subset U$ and $B_\delta(x)$ has compact closure (where we use local compactness of X). By compactness of A we may extract from $\mathcal{O}$ a finite subcover $\mathcal{O}_N = \{B_{\delta_1}(x_1), \dots, B_{\delta_N}(x_N)\}$; we abbreviate each $B_{\delta_i}(x_i)$ to B_i.

Set $V = \bigcup_{i=1}^N B_i$. Obviously $V \subset U$. Now group together intersecting B_i. More specifically, define

$$\begin{aligned}
V_1 &= \bigcup_{j\geq 1,\ B_1\cap B_j\neq\emptyset} B_j, \\
V_2 &= \bigcup_{j\geq 1,\ B_2\cap B_j\neq\emptyset} B_j \setminus V_1, \\
&\vdots \\
V_N &= \bigcup_{j\geq 1,\ B_N\cap B_j\neq\emptyset} B_j \setminus (V_1 \cup \cdots \cup V_{N-1}).
\end{aligned}$$

Then $V_1 \cup \cdots \cup V_N = V$. Let $1 < n \leq N$ be defined by $V_n \neq \emptyset$, $V_{n+1} = \emptyset$ (with the understanding that $n = N$ if $V_N \neq \emptyset$). Then $U \supset V = V_1 \cup \cdots \cup V_n$ and these are pairwise disjoint, open, non-empty and have compact closure, as required. □

We will also need the following closely related proposition.

Proposition 1.3.6 *Let X be a locally connected metric space and $A \subset X$ be a compact set. Then any open neighbourhood U of A contains an open neighbourhood V of A consisting of a finite number of disjoint non-empty connected open sets.*

Proof. The proof is virtually identical to that of Proposition 1.3.5. We merely point out the categorical distinctions: local compactness of X is replaced by local connectedness; the $B_\delta(x)$ are chosen to be connected. The remainder of the construction is identical: we note that, if different B_i have non-empty intersection, they belong to the same connected component. □

Remark 1.3.7 It is clear from the proofs that if X is both locally connected and locally compact, Propositions 1.3.5 and 1.3.6 together imply that any open neighbourhood U of a compact A contains an open neighbourhood $V(A)$ consisting of a finite number of disjoint non-empty connected open sets with compact closure. This fact will be essential in the proof of the main result of Chapter 2. □

We now present two results which provide characterizations of Liapunov stability and of asymptotic stability in terms of dynamical properties of neighbourhoods of the compact invariant set A.

Lemma 1.3.8 (Nature of Liapunov stability) *Let X be a locally compact metric space, $f : X \to X$ a continuous map and A a compact invariant Liapunov-stable set under f. Then any open neighbourhood U of A contains a compact forward-invariant neighbourhood $\overline{W}$ of A.*

Proof. Take an open neighbourhood U of A, which we assume to have compact closure (otherwise just replace it by a smaller one which does, by Proposition 1.3.5). Consider

$$\overline{U}_n = \bigcap_{i=0}^{n} f^{-i}(\overline{U}) = \{x : f^i(x) \in \overline{U},\ 0 \leq i \leq n\}.$$

We have $\overline{U} = \overline{U}_0 \supset \overline{U}_1 \supset \cdots \supset A$ and $f(\overline{U}_n) \subset \overline{U}_{n-1}$. Moreover, all the $\overline{U}_i$ are compact. Setting $\overline{W} = \bigcap_{i=0}^{\infty} f^{-i}(\overline{U})$, it follows that $\overline{W}$ is compact and $f(\overline{W}) \subset \overline{W}$.

By Liapunov stability we know that there is an open neighbourhood V of A such that $f^n(V) \subset U$ for all $n \geq 0$, or equivalently $V \subset f^{-n}(U)$ for all $n \geq 0$. Therefore $V \subset \bigcap_{i=0}^{\infty} f^{-i}(U) \subset \bigcap_{i=0}^{\infty} f^{-i}(\overline{U}) = \overline{W}$. Thus $\overline{W}$ contains V and is therefore a compact neighbourhood of A. □

Lemma 1.3.9 (Nature of asymptotic stability) *Let X be a locally compact metric space, $f : X \to X$ a continuous map and A a compact invariant set. Then A is asymptotically stable if and only if it has a basis of compact neighbourhoods $\{\overline{W}_\alpha\}_{\alpha \in I}$ such that $f(\overline{W}_\alpha) \subset \overline{W}_\alpha$ and $\bigcap_{n=0}^{\infty} f^n(\overline{W}_\alpha) = A$.*

Before proving this lemma it is useful to show the equivalence of the definition of asymptotic stability 1.2.10 – which is the standard in topological dynamics – to another, more geometric, characterization.

Proposition 1.3.10 *Let X be a locally compact metric space, $f : X \to X$ a continuous map, and A a compact invariant set. The following conditions are equivalent:*

(AS1) *A is asymptotically stable.*
(AS2) *For any sufficiently small neighbourhood U of A the successive images $f^n(U)$ converge to A in the sense that for any neighbourhood V of A there exists n_0 such that $f^n(U) \subset V$ for all $n \geq n_0$.*

Proof. We first prove that (AS2) implies asymptotic stability. Given any open neighbourhood $\hat{U}$ of A, choose an open neighbourhood $U \subset \hat{U}$ which satisfies (AS2). Fix an open neighbourhood $V \subset U$, determining a corresponding n_0 as in (AS2). Define $W = \bigcap_{i=0}^{n_0} f^{-i}(V)$. Then W is an open neighbourhood of A; for $0 \leq i \leq n_0$, $f^i(W) \subset V \subset U$; and as $W \subset V \subset U$, we have by (AS2) that $f^i(W) \subset V$ for $i \geq n_0$. So given $\hat{U}$, W is an open neighbourhood of A such that $f^n(W) \subset U$ for all n – proving Liapunov stability of A. To see that $\omega(x) \subset A$ for all $x \in U$ we argue in the following way. Let $x \in U$ and suppose $y \in \omega(x)$ with $y \notin A$. By normality (see Simmons [110]) of X we may choose disjoint open neighbourhoods $V(A)$, $V(y)$ of A and y respectively. By (2) there exists n_0 such that $f^n(U) \subset V(A)$ for all $n \geq n_0$. In particular $f^n(x) \notin V(y)$ for $n \geq n_0$ and therefore $y \notin \omega(x)$, contradicting our assumption. Thus for all $x \in U$ we have that $\omega(x) \subset A$ – proving asymptotic stability of A.

To show that (AS1) implies (AS2) we choose a neighbourhood U of A with the following properties:

1. U has compact closure,
2. $\forall x \in \overline{U}, \quad \omega(x) \subset A$.

Condition (1) may be met as a consequence of Proposition 1.3.6, while condition (2) is a consequence of asymptotic stability, by eventually shrinking U.

We first show that, given any open neighbourhood V of A with $V \subset U$, there exists $N = N(V)$ such that $n \geq N \Rightarrow f^n(\overline{U}) \subset V$. This is a two-step procedure. First we show that, given U, V as above, for all $x \in \overline{U}$ there exists $n_0(x)$ such that $n \geq n_0(x)$ implies $f^n(x) \in V$. Suppose not. Then there are $x \in \overline{U}$, $n_k \to \infty$ such that $f^{n_k}(x) \in \overline{U} \setminus V$. As this is compact, there is a convergent subsequence $f^{n_{k_j}}(x) \to y \in \overline{U} \setminus V$. But $y \notin A$ and $y \in \omega(x)$, contradicting property (2) of U.

Next we show that $n_0(x)$ may be taken uniformly in $\overline{U}$. Let U, V be as before and $W \subset V$ be an open neighbourhood of A satisfying the Liapunov stability condition with respect to V, that is, $f^n(W) \subset V$ for all $n > 0$. This W satisfies the conditions of the previous step; that is, for all $x \in \overline{U}$ there exists $n_0(x)$ such that $n \geq n_0(x) \Rightarrow f^n(x) \in W$. By continuity there is an open ball $B_{\epsilon_x}(x)$ around x such that $y \in B_{\epsilon_x}(x) \Rightarrow f^{n_0(x)}(y) \in W$. Then the collection $\mathcal{B} = \{B_{\epsilon_x}(x)\}_{x \in \overline{U}}$ is an open cover of $\overline{U}$, from which by compactness we can extract a finite subcover $\mathcal{B}' = \{B_{\epsilon_1}(x_1), \ldots, B_{\epsilon_k}(x_k)\}$. Set $N = \max\{n_0(x_1), \ldots, n_0(x_k)\}$. Then for every $B_{\epsilon_i}(x_i)$, $i = 1, \ldots, k$ there is an integer j between 0 and N such that $f^j(B_{\epsilon_i}(x_i)) \subset W$. By the choice of W, this implies that $f^n(B_{\epsilon_i}(x_i)) \subset V$ for all $n \geq N$, $i = 1, \ldots, k$. As $\overline{U} \subset B_{\epsilon_1}(x_1) \cup \cdots \cup B_{\epsilon_k}(x_k)$, it follows that $n \geq N \Rightarrow f^n(\overline{U}) \subset V$. □

Proof of Lemma 1.3.9.
1. *Proof of 'only if' part.* Suppose A is asymptotically stable. By Proposition 1.3.10 we may choose U as in (AS2). Suppose there exists $x \in \Lambda = \bigcap_{n \geq 0} f^n(U)$ with $x \notin A$. By normality and local compactness of X we may choose disjoint open neighbourhoods $V(A)$, $V(x)$ with compact closure contained in U. By (AS2) we know that, for $n \geq n_0$, $f^n(U) \subset V(A)$; in particular $x \notin f^n(U)$ for all $n \geq n_0$, contradicting the hypothesis. Therefore $\bigcap_{n=0}^{\infty} f^n(U) = A$.

To see that there exists a basis of compact neighbourhoods $\overline{W}_\alpha$ with the claimed properties, fix an arbitrary neighbourhood $V_\alpha \subset U$ with compact closure and set $\overline{W}_\alpha = \bigcap_{i \geq 0} f^{-i}(\overline{V}_\alpha)$. $\overline{W}_\alpha$ is compact and $f(\overline{W}_\alpha) \subset \overline{W}_\alpha$. By Proposition 1.3.10 there exists n_0 such that $n \geq n_0$ implies $f^n(U) \subset V$; in particular $f^n(\overline{V}) \subset V$. Therefore $\overline{V} \subset f^{-n}(V)$ for $n \geq n_0$. So in fact $\overline{W}_\alpha = \bigcap_{i=0}^{\infty} f^{-i}(\overline{V})$, whereby $\overline{W}_\alpha$ is a neighbourhood of A, completing the proof of the first implication.

2. *Proof of 'if' part:* trivial. □

Remark 1.3.11 In fact, both conditions in Lemma 1.3.9 are equivalent to the following, apparently weaker, condition: for any sufficiently small neighbourhood of U of A, $\bigcap_{n=0}^{\infty} f^n(U) = A$ – as shown, for instance, by Milnor [70]. It is easy to see that this condition implies the existence of a basis of neighbourhoods as stated in Lemma 1.3.9. So asymptotic stability is equivalent to this single condition, which is usually treated in the literature as an independent definition of 'attractor' (see, e.g., Smale [113]) or, more precisely, of 'attracting set' – see § 1.6. □

Remark 1.3.12 If A is a compact, invariant, Liapunov-stable but *not* asymptotically stable set, then A must be accumulated by ω-limit sets other than itself – in other words, in any neighbourhood of A there are infinitely many ω-limit sets distinct from A. The most direct way of proving this fact is probably the following. By Lemma 1.3.8 every neighbourhood of A contains a compact neighbourhood $\overline{W}$ with $f(\overline{W}) \subset \overline{W}$. Remark 1.2.5 then tells us that, for all $x \in \overline{W}$, $\omega(x) \subset \overline{W}$. On the other hand, there is at least one $x \in \overline{W}$ such that $\omega(x) \not\subset A$, as otherwise A would be asymptotically stable. As both A and $\omega(x)$ are compact invariant, this means that $\omega(x) \cap A = \emptyset$. Arbitrariness of $\overline{W}$ implies the result. □

We now study connectivity properties of compact invariant sets and their relation with dynamics. In doing so we set up a construction which will be essential for Chapter 2. The next results deal with different but related aspects of this question.

Let X be a locally compact metric space, and let $f : X \to X$ be a continuous map. Suppose $A \subset X$ is a compact invariant set. Let $\sim$ be the equivalence relation on A determined by its connected components; that is, given $x, x' \in A$, then $x \sim x'$ if and only if x and x' lie in the same component of A. Let $K = A/\sim$ with the identification topology. Then the diagram

$$\begin{array}{ccc} A & \xrightarrow{f} & A \\ \pi\downarrow & & \downarrow\pi \\ K & \xrightarrow{\tilde{f}} & K \end{array}$$

commutes, where π is the identification map and $\tilde{f}$ is the map induced by f.

We use the notations K, π, $\tilde{f}$ for this construction throughout Chapters 1 and 2 when no ambiguity arises. We occasionally refer to K as the *space of connected components* of A. The following result characterizes the topology of the space of connected components of an invariant set as well as the projection of the dynamics on it. We restate the standing assumptions as this basic result will be essential throughout.

Lemma 1.3.13 *Let X be a locally compact metric space, $A \subset X$ a compact non-empty set, $\sim$ the quotient by connected components, $K = A/\sim$ the identification space and π the identification map. Then K is metrizable, and the identification topology on K coincides with the topology induced by the Hausdorff distance between components of A.*

Proof. Let ρ be the metric on X. Let $A = \bigcup_{\alpha \in K} C_\alpha$, where the $C_\alpha = \pi^{-1}(\alpha)$ are the connected components of A. Let $\mathcal{S} = \{C_\alpha : \alpha \in K\}$. Then $\mathcal{S}$ is a metric space with the Hausdorff metric

$$d_H(C_\alpha, C_\beta) = \inf_{\delta > 0}\{\delta : C_\alpha \subseteq C_\beta^{(\delta)} \text{ and } C_\beta \subseteq C_\alpha^{(\delta)}\},$$

where $C^{(\delta)}$ is the δ-neighbourhood of C. Give K the induced metric

$$\tilde{d}(\alpha, \beta) = d_H(C_\alpha, C_\beta).$$

We thus have two ways to topologize K: with the identification topology or the $\tilde{d}$-topology. To prove that they coincide, we are required to show that a set $G \subseteq K$ is open with respect the the identification topology if and only if it is open in the $\tilde{d}$-topology. In what follows, we shall abbreviate these phrases to *i-open* and *$\tilde{d}$-open.*

Suppose that $G \subseteq K$ is $\tilde{d}$-open. Then for all $\alpha \in G$ there exists $\delta > 0$ such that $\tilde{B}_\delta(\alpha) \subset G$, where $\tilde{B}_\delta(\alpha) = \{\beta : \tilde{d}(\alpha, \beta) < \delta\}$. By definition of π, we have

$$\pi^{-1}(G) = \{C_\gamma : \gamma \in G\}$$

and

$$\pi^{-1}(\tilde{B}_\delta(\alpha)) = \{C_\beta : \tilde{d}(\alpha, \beta) < \delta\} = \{C_\beta : d(C_\alpha, C_\beta) < \delta\}.$$

Hence

$$\pi^{-1}(\tilde{B}_\delta(\alpha)) \subset \pi^{-1}(G).$$

Therefore $\pi^{-1}(G)$ is open, so G is i-open.

Conversely, suppose that G is i-open. Then $\pi^{-1}(G)$ is open in the relative topology of A, that is, $\pi^{-1}(G) = G \cap U$ for some open $U \subset X$. Let $\alpha \in G$; then $C_\alpha \stackrel{\text{def}}{=} \pi^{-1}(\alpha) \subset \pi^{-1}(G)$. Now, C_α is compact and contained in U; there is a Lebesgue number $\delta > 0$ such that $C_\alpha^{(\delta)} = \{x \in X : \rho(x, C_\alpha) < \delta\}$, the δ-neighbourhood of C_α, is contained in U. Choosing $\epsilon < \delta$, it follows that $d_H(C_\alpha, C_\beta) < \epsilon \Rightarrow C_\beta \subset C_\alpha \subset U$, or $d_H(C_\alpha, C_\beta) < \epsilon \Rightarrow C_\beta \subset \pi^{-1}(G)$ (because $\pi^{-1}(G) = U \cap A$ and $C_\beta \subset A$). Equivalently, $\tilde{d}(\alpha, \beta) < \epsilon \Rightarrow \beta \in G$, and therefore G is $\tilde{d}$-open.

Thus the identity map from the metric space $(K, \tilde{d})$ to the topological space (K, i) is a homeomorphism. This shows the latter to be metrizable and the metric $\tilde{d}$ to be compatible with its topology. □

If in addition to the above assumptions A is topologically transitive, then we have the following characterization theorem for the space of connected components.

Theorem 1.3.14 *Let X be a locally compact metric space, $f : X \to X$ a continuous map and A be a compact transitive set under f. Let K be the space of components of A and $\tilde{f} : K \to K$ the induced map. Then one of the following holds.*

1. *K is finite and $\tilde{f}$ is a cyclic permutation of K.*
2. *K is a Cantor set and $\tilde{f}$ is transitive on K.*

Proof. K is by definition totally disconnected, and is compact by compactness of A and continuity of π. If x is a transitive point in A, then $\pi(x)$ is a transitive point for $\tilde{f}$ in K. Thus transitivity of A under f implies transitivity of K under $\tilde{f}$ irrespective of the cardinality of K.

If K is finite it is homeomorphic to $\{1, \ldots, k\}$ with the discrete topology. Transitivity of $\tilde{f}$ then implies it is a cyclic permutation on K.

If K is infinite, transitivity of $\tilde{f}$ implies that K is perfect. Thus K is totally disconnected, compact, perfect and metric, and therefore homeomorphic to the Cantor set by a standard result in point-set topology; see e.g. Hocking and Young [48], Corollary 2-98. □

Remark 1.3.15 Theorem 1.3.14 assumes no connectedness properties of the underlying space X. In the special case where A is the Cantor set, we see that Theorem 1.3.14 is, in particular, a statement about arbitrary transitive maps of the Cantor set. In particular all transitive maps of sequence spaces are included.

At this level of generality, no stronger statements about $\tilde{f}$ are possible. This is in sharp contrast with the situation where X is locally connected and A is Liapunov stable, as we will see in Chapter 2; these apparently mild conditions impose strong restrictions on the dynamics of $\tilde{f}$. □

We now consider a slightly different aspect of the dynamics on components. The transitive sets which arise naturally in dynamics are most often nowhere dense in the ambient space (even though they may or may not have positive Lebesgue measure; see Bowen [16]). We prove below a result for transitive sets with non-empty interior.

Proposition 1.3.16 *Let X be a locally connected metric space and $f:X \to X$ be a continuous map. Suppose that A is a transitive set such that* $\operatorname{int} A \neq \emptyset$. *Then A has a finite number of connected components $A_0, A_1, \ldots, A_{k-1}$ which are cyclically permuted by f. If X is a C^r manifold and f is a C^s map, $1 \leq s \leq r$, then in addition* $\operatorname{int} A_j \neq \emptyset$ *for $j = 0, \ldots, k-1$.*

Proof. Since $\operatorname{int} A \neq \emptyset$ and X is locally connected, we may choose $x \in A$ and $\delta > 0$ such that $B_\delta(x) \subseteq \operatorname{int} A$ and $B_\delta(x)$ is connected. Then $B_\delta(x) \subseteq C_x$. By Lemma 1.3.1 transitive points are dense in A. Thus there exists a transitive point $y \in B_\delta(x)$ and an integer $n > 0$ such that $f^n(y) \in B_\delta(x)$. Denoting by C_y the connected component of A containing y, this shows that $f^n(C_x) = C_x$ (equality coming from transitivity).

With the notation of Theorem 1.3.14, it follows that $\tilde{f}^n : K \to K$ is the identity map. Thus each orbit in K is periodic, and transitivity of $\tilde{f}$ implies K is finite; $\tilde{f}$ is thus a cyclic permutation by Theorem 1.3.14, proving the first statement.

For the second statement, assume X is a differentiable manifold and f is a differentiable map. The argument above applies and $A = A_0 \cup A_1 \cup \cdots \cup A_{k-1}$, where the A_j are the connected components of A. Without loss of generality assume $\operatorname{int} A_0 \neq \emptyset$. Denote the dimension of the manifold by n and the derivative of f at the point p by $d_p f : T_p A \to T_{f(p)} A$.

Suppose for a contradiction that rank $d_p f \leq n-1$ for all $p \in \operatorname{int} A_0$. Then, by continuity of the derivative, rank $d_p f \leq n-1$ for all $p \in A_0$, and hence every $p \in A_0$ is a critical point of f. By Sard's Theorem, $f(A_0)$ has zero Lebesgue measure; since f is C^1 this implies that, for all $n > 0, f^n(A_0)$ has zero measure and consequently empty interior. But then $f^k(A_0) = A_0$ has empty interior, contradicting our assumption.

Thus there exists $p_0 \in \operatorname{int} A_0$ such that rank $d_p f = n$. Since f is C^1 there is a neighbourhood U_0 of p_0 such that for all $p \in U_0$, rank $d_p f = n$; hence by the inverse function theorem $f_{|U_0}$ is a local diffeomorphism onto its image. In particular, $f(U_0) \subseteq A_1$ is open in X, and A_1 has non-empty interior. Applying this argument inductively shows that every A_j has non-empty interior. □

1.4 Attractors

The stability conditions introduced in § 1.3 are not enough to characterize a compact invariant set as an 'attractor' from a topological point of view. To cite a simple example, if f is the identity map on a compact manifold X, then any closed subset of X is Liapunov stable and X itself is asymptotically stable.

The above example is enlightening in the following sense. We would like to impose conditions on our definitions of attractor in order to rule out the possibility of the dynamics inside an attractor A being trivial. We thus have, from a programmatic point of view, two distinct but essential conditions that a reasonable definition of attractor should satisfy. Firstly, an "attractivity" condition, whose prototype in the topological category is asymptotic stability. Secondly, an "irreducibility" condition – transitivity in the topological category – which ensures that every piece of an attractor plays an essential role.

Different formal conditions imposed to realize this programme give rise to distinct definitions of attractor present in the literature. As we have stated, all the definitions we are about to introduce have their degree of usefulness, as they illuminate different aspects of the dynamics. Examples of the distinctions between the definitions introduced below are given in § 1.5.

Definition 1.4.1 (Axiom A attractors) *Let X be a smooth compact manifold and $f : X \to X$ a diffeomorphism. A compact invariant set A is an* Axiom A attractor *if:*

1. *A admits a hyperbolic structure;*
2. *periodic points are dense in A;*
3. *A is topologically transitive under f;*
4. *there exists a neighbourhood $U(A)$ such that $\bigcap_{n \geq 0} f^n(U) = A$.*

Remark 1.4.2 This definition is originally due to Smale [113]. Saying that A admits a hyperbolic structure means that there exists a continuous splitting of the tangent bundle $T_A X$, invariant under df, into expanding and contracting directions. Compactness of A implies that this splitting is uniform, so that A is sometimes said to be "uniformly hyperbolic". Uniform hyperbolicity has an enormous wealth of implications for dynamics; see § 1.6 for a brief discussion of these. □

Remark 1.4.3 It is possible to extend this definition to the class of smooth maps, not necessarily invertible, on smooth compact manifolds. This requires several redefinitions. The invariant manifolds cannot be defined and constructed as in the invertible case; in particular the global stable 'manifold' need not be a submanifold, see Shub [107]. However, a suitable form of hyperbolicity is still possible, see Palis and Takens [86]; it implies the existence of local stable and unstable invariant manifolds (Irwin's proof of the local stable manifold theorem [50] and the graph transform proof of the local unstable manifold theorem [108] only require the forward iterates of the map) and the fact that $A = \bigcup_{x \in A} \overline{W^u(x)}$. So even in this context Definition 1.4.1 is meaningful. □

This is certainly the strongest definition of attractor to be found in the dynamical systems literature. Several related comments are to be found in § 1.6.

The next definition which we introduce is closely related to the previous one; essentially it relaxes the conditions of A having a hyperbolic structure (so that it makes sense in a non-differentiable setting) and a dense set of periodic points.

Definition 1.4.4 (Asymptotically Stable Attractors) *Let X be a metric space and $f : X \to X$ a continuous map. A compact invariant set A is said to be an* asymptotically stable attractor *if*

1. *A is asymptotically stable;*
2. *A is topologically transitive.*

Remark 1.4.5 As described in some detail in §1.6, this definition has its origins in topological dynamics. Asymptotically stable attractors are sometimes referred to in the dynamical systems literature as 'open-basin attractors'. In fact, this terminology is misleading, as explained in § 1.6. In § 1.5 we will present examples of 'attractors' (in some, very reasonable, sense) which have an open basin of attraction but are *not* asymptotically stable. □

It is clear from Lemma 1.3.9 that Axiom A attractors are asymptotically stable attractors.

We next provide an interesting application of the ideas in § 1.3.

Theorem 1.4.6 *Let X be a locally compact, locally connected metric space and $f : X \to X$ a continuous map. Suppose A is an asymptotically stable attractor for f. Then A has finitely many connected components which are cyclically permuted.*

Proof. By Propositions 1.3.5 and 1.3.6, Remark 1.3.7 and Lemma 1.3.9, we may construct an open neighbourhood W of A with compact closure such that

1. W has finitely many connected components $W_1, \ldots, W_n$ with disjoint compact closure;
2. $f(\overline{W}) \subset \overline{W}$;
3. $\bigcap_{n=0}^{\infty} f^n(\overline{W}) = A$.

Transitivity of A and connectedness of the W_i imply that we can renumber the W_i in such a way that $f(\overline{W}_i) \subset \overline{W}_{i+1 \pmod n}$. Finally, the fact that each W_i contains exactly one connected component of A stems from Lemma 1.3.9, as $A_i = \bigcap_{k=0}^{\infty} f^{kn}(\overline{W}_i)$ is connected. Therefore A has exactly n connected components $A_1, \ldots, A_n$ which satisfy $f(A_i) = A_{i+1 \pmod n}$. □

The next two definitions are due to Milnor [69] and blend topological and measure-theoretical concepts. The relevant structure is the following. X is a compact, or at least finite-dimensional, smooth manifold, $f : X \to X$ is a continuous map and μ is a measure which is equivalent to Lebesgue measure on X. Here equivalent means having the same null sets. Lebesgue measure may be derived, for instance, from a volume form associated with a Riemannian structure on X.

Definition 1.4.7 (Milnor Attractors) *Let X be a finite-dimensional manifold, $f : X \to X$ be a continuous map and μ be a Borel measure equivalent to Lebesgue measure. A compact invariant set A is said to be a* Milnor attractor *if*

1. *its basin of attraction $\mathcal{B}(A)$ has positive μ-measure;*
2. *there is no proper closed subset $A' \subset A$ such that $\mathcal{B}(A)$ and $\mathcal{B}(A')$ differ by a set of zero μ-measure.*

Definition 1.4.8 (Minimal Milnor Attractors) *Let X be a finite-dimensional smooth manifold, $f : X \to X$ be a continuous map and μ be a Borel measure equivalent to Lebesgue measure. A compact invariant set A is said to be a* minimal Milnor attractor *if*

1. *its basin of attraction $\mathcal{B}(A)$ has positive μ-measure;*
2. *there is no proper closed subset $A' \subset A$ for which $\mathcal{B}(A')$ has positive μ-measure.*

Remark 1.4.9 These definitions are clearly independent of the choice of the measure μ. Indeed, μ is only used to distinguish between positive and zero-measure sets; as this distinction is invariant under the relation of measure-theoretical equivalence, the independence of μ follows. □

Remark 1.4.10 The term *minimal* in "minimal Milnor attractor" has no relation with topological minimality introduced in Definition 1.2.7; see Remark 1.2.8. This should not be a great source of confusion. We use minimality in the sense of Definition 1.4.8 exclusively within the phrase "minimal Milnor attractors"; if while considering minimal Milnor attractors the need arises to discuss minimality in the sense of Definition 1.2.7, we will refer to it specifically as *topological minimality*. In all other contexts reference to minimality will mean topological minimality by default.

Although this duplication of meanings may be cumbersome, any alternative other than introducing even more terminology would lead to mathematical rather than linguistic difficulties; see § 1.6. □

Remark 1.4.11 If A is a Milnor attractor it does not follow that there exists $x \in \mathcal{B}(A)$ such that $\omega(x) = A$. In particular Milnor attractors are not necessarily transitive. Examples are easily provided: this happens for instance when A is the union of several disjoint asymptotically stable sets or attractors, the simplest case being that where A is the set consisting of two asymptotically stable fixed points. This extreme form of decomposability is not allowed for *minimal* Milnor attractors. In this case a direct consequence of condition (2) in Definition 1.4.8 is that $\omega(x) = A$ for (Lebesgue) a.e. $x \in \mathcal{B}(A)$. However, even minimal Milnor attractors are not necessarily transitive, see Example 1.5.6. This is not surprising, since the indecomposability condition (2) in Definition 1.4.8 is not purely topological but also measure-theoretical. □

Remark 1.4.12 It might be expected that asymptotically stable attractors would automatically be Milnor attractors. In fact, the motivation for the definition of Milnor attractors was to generalize the concept of basin of attraction from a topological to a measure-theoretical setting.

The situation, however, is more subtle than might appear. The basin of an asymptotically stable attractor is open and thus has positive measure, see Hirsch [47]. Condition (1) in Definition 1.4.7 is thus satisfied. However, it is at present unknown whether topological transitivity implies condition (2). The reason is that these definitions deal respectively with the topological and measure-theoretical categories. However, the dual concepts of Baire category and Lebesgue measure are to a large extent independent, see Oxtoby [84]; it is a standard matter to construct second category sets whose complement has zero Lebesgue measure. For an example in $\mathbb{R}$ see Royden [98].

It may seem surprising that asymptotically stable attractors might not be Milnor attractors; a counterexample would probably be pathological. On the other hand, the above considerations allow us to show that asymptotically stable attractors are *not* necessarily minimal Milnor attractors. We construct one such example in 1.5.5. The point at hand is that both definitions imply that closed invariant proper subsets of an attractor A are "small". In the topological sense, this means by Corollary 1.3.2 that they are nowhere dense in A. In the measure-theoretical sense, this means that their basin has zero Lebesgue measure. For further comments see § 1.6. □

Remark 1.4.13 Milnor attractors do *not* necessarily have any stability properties from the topological point of view. They can clearly be Liapunov-unstable sets; for instance, if there is a zero-measure family of local unstable manifolds emanating from, but not contained, in A, it will be an unstable Milnor attractor. A simple example is given in § 1.5 below.

The following definition does not really introduce a new concept of attractor; it is rather a specialization of Milnor's definition to a case of particular interest.

Definition 1.4.14 *A compact invariant set A is an* essentially asymptotically stable *or* Melbourne *attractor if there exists a set C such that, given any open neighbourhood U of A and any $\alpha \in (0,1)$, there is an open neighbourhood $V \subset U$ of A such that*

1. *all orbits starting in $V \setminus C$ remain in U and are asymptotic to A,*
2. *$\ell(V \setminus C)/\ell(V) > \alpha$, where ℓ is Lebesgue measure.*

Given a Milnor attractor A for a continuous map $f : X \to X$ of a compact manifold X, consider the Birkhoff average of a continuous function $\phi : X \to X$ on the orbit starting at x:

$$L(\phi, x) = \lim_{n\to\infty} \frac{1}{n} \sum_{i=0}^{n-1} \phi(f^i(x)).$$

Suppose this limit exists and is independent of x a.e. in $\mathcal{B}(A)$ or, at least, in a subset $B \subset \mathcal{B}(A)$ of positive Lebesgue measure (this is often assumed in physics, where it is taken for granted that a randomly chosen point will display the 'typical' ergodic behaviour with positive probability). Then L is a continuous linear functional in $C(X, \mathbb{R})$. Furthermore, it is positive and normalized. By the Riesz representation Theorem [65] it defines a unique probability measure μ such that

$$\lim_{n\to\infty} \frac{1}{n} \sum_{i=0}^{n-1} \phi(f^i(x)) = \int_A \phi d\mu \tag{1.2}$$

for all $\phi \in C(X, \mathbb{R})$ and a.a. $x \in B$ (it is clear that μ is invariant and supported in A). This is the basis of the following definition, in which the noun 'attractor' refers both to the case where A is a Milnor attractor as to the case where it is an asymptotically stable attractor.

Definition 1.4.15 (SBR measure and attractor) *Let X be a locally compact manifold, $f : X \to X$ a continuous map admitting an attractor A. If there exists an ergodic measure μ supported in A such that (1.2) is satisfied for all continuous ϕ and a.a. $x \in B$, where B has positive Lebesgue measure, then μ is called an* SBR measure *and A an* SBR attractor.

We know by Remark 1.2.15 that any Milnor attractor supports at least one ergodic measure. Furthermore, Birkhoff's ergodic theorem (see Mañé [65]) states that (1.2) is satisfied almost everywhere *with respect to μ*. However, in most cases A has zero Lebesgue measure; thus even if an SBR measure exists it is singular with respect to Lebesgue measure.

The existence of an SBR measure for an attractor is a difficult mathematical problem. As opposed to ergodic measures, there are no general results on the existence of SBR measures. The most general instance in which their existence can be proved are that of Axiom A attractors, see Ruelle [100], and non-uniformly hyperbolic attractors, see Pugh and Shub [97]. We sketch these below.

Let X be a Riemannian manifold and $f : X \to X$ be a $C^{1+\alpha}$ map. Suppose A is an asymptotically stable attractor under f on which a hyperbolic or a non-uniformly hyperbolic structure is defined.

Definition 1.4.16 (strong SBR measure) *A* strong SBR measure *for A is an ergodic invariant measure μ whose conditional measures μ_σ on unstable manifolds W_σ in A are absolutely continuous with respect to the Riemannian measure induced on these manifolds.*

Theorem 1.4.17 *Strong SBR measures are SBR measures. Moreover, in the Axiom A case the complement of B in $\mathcal{B}(A)$ has zero Lebesgue measure.*

Proof. See Ruelle [100] for the Axiom A case, Pugh and Shub [97] for the non-uniformly hyperbolic case. □

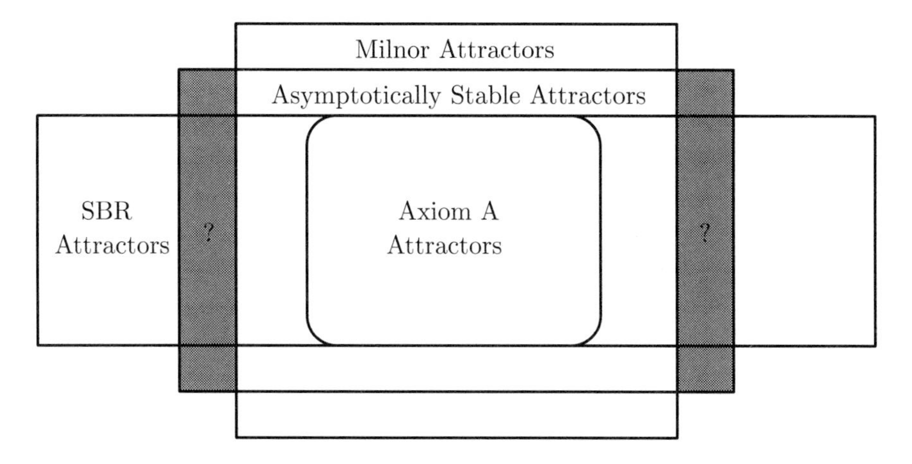

Figure 1.1: Schematic diagram of the hierarchy of the definitions of attractor.

The only general method for proving that a measure is SBR is by disintegrating it along unstable manifolds and showing this disintegration to be absolutely continuous with respect to the Riemannian measure induced on unstable manifolds – in other words, showing it to be strongly SBR. For a discussion see § 1.6.

We summarize the present discussion of attractors in Figure 1.1.

1.5 Examples and counterexamples

This section consists of examples whose purpose is either to stress the optimal status of a theoretical point discussed above or to illustrate the fine points of the distinctions introduced in § 1.4. Our first result concerns Proposition 1.3.16.

Example 1.5.1 The differentiability of f is essential for the second conclusion in Proposition 1.3.16 – only in the special case $X = \mathbb{R}$ is it superfluous, as non-trivial components in $\mathbb{R}$ automatically have non-empty interior. We will construct a geometrically simple counterexample in $\mathbb{R}^2$ such that (i) f is continuous; (ii) A is a transitive set with non-empty interior; (iii) $A = A_0 \cup A_1$ with A_0, A_1 components of A, int $A_0 = \emptyset$ and int $A_1 \neq \emptyset$.

Let $X = \mathbb{R}^2$, $A_0 = [0,1] \times \{0\}$, $A_1 = [0,1] \times [1,2]$. Define $\varphi_1 : A_1 \to A_0$ by $\varphi_1(x, y) = (x, 0)$. That is, φ_1 is just projection onto the x-axis. The main step is to define $\varphi_0 : A_0 \to A_1$; it then follows from Urysohn's Lemma that there exists a continuous $f : \mathbb{R}^2 \to \mathbb{R}^2$ such that $f_{|A_0} = \varphi_0, f_{|A_1} = \varphi_1$. In fact, we can even extend the φ's so that $A = A_0 \cup A_1$ becomes an asymptotically stable attractor.

The map φ_0 is a Peano curve, defined so that $f^2_{|A_0} = \varphi_1 \circ \varphi_0 : A_0 \to A_0$ is a transitive map of the interval A_0. If we can construct such a map, then we are finished; for if $z = (x, 0) \in A_0$ is such that $\omega(z) = A_0$ under f^2, then obviously

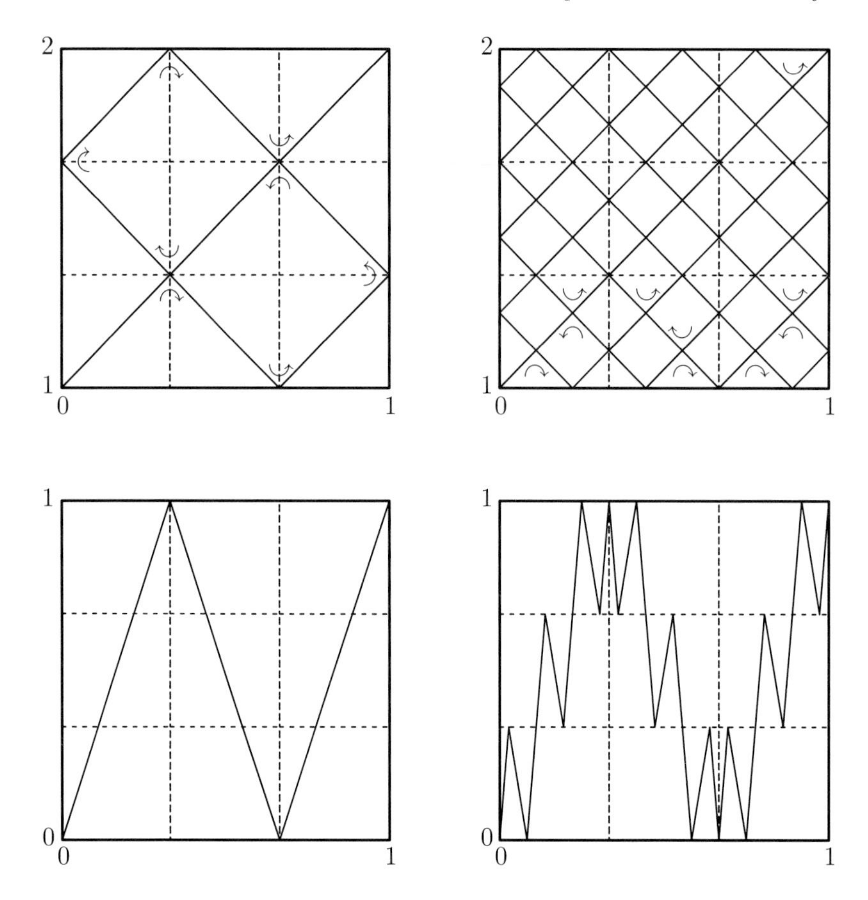

Figure 1.2: First steps in the construction of φ_0 and x_0.

$\omega(f(z)) = f(A_0) = A_1$ under f^2, so $\omega(z) = \omega(z, f^2) \cup \omega(f(z), f^2) = A_0 \cup A_1$ under f. Therefore z is a transitive point for f.

It remains to construct φ_0. We give only a geometrical construction; specific formulae may be derived if so desired. We use a suitable adaptation of Sierpiński's example of a Peano curve, as described in Kuratowski [58]. Denote by $I = [0, 1]$ the unit interval, $Q = [0, 1] \times [1, 2]$ the translated unit square in the plane; construct a sequence of functions $\varphi^{(n)} = (x^{(n)}(t), y^{(n)}(t))$, $\varphi^{(n)} : I \to Q$ as indicated in Figure 1.2.

It is easy to check that $\|\varphi^{(n)} - \varphi^{(n+1)}\|_0 \leq 1/3^n$, so the $\varphi^{(n)}$ converge uniformly to a continuous $\varphi_0 : I \to Q$, $\varphi_0 = (x_0, y_0)$. On the other hand, every $x^{(n)}$ maps each of $[0, 1/3]$, $[1/3, 2/3]$, $[2/3, 1]$ *onto* $[0, 1]$; the same is true for x_0.

Consider the map $x_0 : I \to I$. Since $f^2_{|A_0}(t, 0) = \varphi_1 \circ \varphi_0(t, 0) = (x_0(t), 0)$, it is enough to prove that $x_0 : I \to I$ is topologically transitive. The last observation in

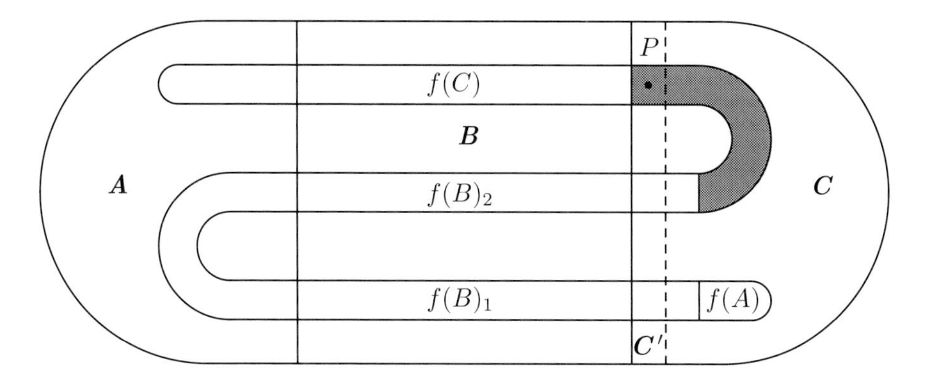

Figure 1.3: A modified horseshoe map.

the previous paragraph implies that any open subinterval $J \subset I$ is mapped *onto* [0,1] after a finite number of iterations of x_0, since such an interval always contains a subinterval of the form $J_{k,m} = [k/3^m, (k+1)/3^m]$ for certain integers k, m and $x_0^m(J_{k,m}) = [0,1]$. This condition implies that $x_0 : [0,1] \to [0,1]$ is topologically mixing, hence transitive; see Block and Coppel [14].

Thus $A = A_0 \cup A_1$ is a transitive set under f with non-empty interior, having two components, one of them with empty interior. This situation is possible only because f fails to be C^1.

This example may be adapted to produce mappings $f_n : \mathbb{R}^m \to \mathbb{R}^m$ with transitive sets $A^{(n)} = C_0 \cup \cdots \cup C_{n-1}$, where the C_j are the connected components of $A^{(n)}$, such that each component has arbitrary (integral) dimension $\leq m$. These components are cycled by Proposition 1.3.16, so that $f_n(C_i) = C_{i+1 \pmod n}$; but we can construct f_n in such a way that $d_i = \dim C_i$ is an arbitrary positive integer less than or equal to m. □

Our second example shows that Theorem 1.3.14 ceases to be valid if we replace transitivity of A by the weaker condition of A being an ω-limit set.

Example 1.5.2 The following example is based in a generalized Smale horseshoe construction [113]. Let D^2 be the compact 2-disk. Construct an injective differentiable map $f : D^2 \to D^2$ in the way sketched in Figure 1.3, where for simplicity we assume that the usual hyperbolicity conditions are satisfied. We extend f to a diffeomorphism $\tilde{f} : S^2 \to S^2$ of the 2-sphere by first extending f to $\mathbb{R}^2$ and then performing one-point compactification by adding a (hyperbolic) repellor at infinity [113]. $f(B)$ is constructed in the usual way (see Smale [113]); there is a hyperbolic Cantor set $K \subset B$ invariant under f and f^{-1} and restricted to which the dynamics is conjugate to a full 2-shift. The stable manifold of this hyperbolic set is the vertical bundle $K \times V$ and the unstable manifold is the horizontal bundle $K \times H$.

The main difference with respect to the classical horseshoe is the definition of $f(C)$. The strip $C' \subset C$ is mapped to the shaded region in such a way that it has a unique hyperbolic fixed point P, which is a saddle whose (local) stable and unstable manifolds are vertical and horizontal respectively. $C \setminus C'$ is mapped onto a strip intersecting B and A as shown.

Define a map $\pi : D^2 \to \Sigma_3 = \{1, 2, C\}^{\mathbb{Z}}$ by associating to $x \in D^2$ its itinerary $\pi(x) = (i_0, i_1, \dots)$, where

$$i_j = \begin{cases} 1 & \text{if} \quad f^j(x) \in (f(B) \cap B)_1 \\ 2 & \text{if} \quad f^j(x) \in (f(B) \cap B)_2 \\ C & \text{if} \quad f^j(x) \in (f(C) \cap C) \end{cases}$$

where $(f(B) \cap B)_1$, $(f(B) \cap B)_2$ denote respectively the lower and upper components of $(f(B) \cap B)$. Note that π is not defined on the whole of D^2; in particular it is not a homeomorphism. Nevertheless, by compactness and the finite intersection property it is surjective.

Considering the sequence

$$\underline{i} = (1C2CC11CCC12CCCC21CCCCC22CCCCCC\dots)$$

we see that the corresponding $x \in D^2$ has an orbit which is dense in the Cantor set $K = \bigcap_{n \in \mathbb{Z}} f^n(B)$ and also spends arbitrarily long stretches of time in C. As by construction the only non-wandering point in C is the fixed point P, there is a subsequence of $f^n(x)$ converging to P. Thus $\omega(x) = K \cup \{P\}$, which is the union of a Cantor set and an isolated point.

Thus the components of an ω-limit set do not necessarily form a finite or a Cantor set in the absence of topological transitivity. Theorem 1.3.14 does imply, however, that the space of connected components of each transitive piece of an ω-limit set is either finite or a Cantor set. □

The next set of examples deal essentially with the question of distinguishing the different types of attractor introduced on § 1.4. In some cases it is possible to think of more striking or sophisticated examples. However, the purpose of these examples is merely to illustrate a point. We therefore chose, whenever possible, to present the most elementary example showing the intended behaviour. In fact, all the definitions introduced in § 1.4 are used in later chapters to classify the attractors at hand, so that these will in due course provide us with a more sophisticated and interesting example gallery.

Example 1.5.3 As remarked earlier, conditions (3) and (4) in definition 1.4.1 of Axiom A attractors imply that Axiom A attractors are asymptotically stable attractors. However, the converse is not true. An asymptotically stable attractor can fail to be Axiom A by failing to satisfy condition (1) or (2) (or both!). For an example of a non-hyperbolic asymptotically stable attractor, one merely has to think of a weakly attracting fixed point – that is, a fixed point such that at least one of the eigenvalues of the linear part lies on the unit circle but the nonlinear terms force it to be (non-hyperbolically) asymptotically stable. As an example of the second kind, where periodic points fail to be dense, it is enough to consider a normally hyperbolically attracting invariant circle on which the dynamics is conjugate to an irrational rotation, which has no periodic points. For an explicit example, take $X = \mathbb{R}^2$ and f to be given in polar coordinates by

$$f(r,\theta) = (r + \frac{\epsilon}{2\pi}\sin(2\pi r), \theta + \alpha) \tag{1.3}$$

where $0 < \epsilon < 1$ and $\alpha \in \mathbb{R} \setminus \mathbb{Q}$. Then the circle $C = \{(r,\theta) : r = 1\}$ is a normally hyperbolically attracting invariant circle, therefore asymptotically stable. $f_{|C}$ is an irrational rotation, which is minimal [3]. It therefore has no periodic points, so that condition (2) is violated.

Of course, it is possible to combine these two examples and produce examples which are neither hyperbolic nor satisfy the "closing lemma" (2). □

The following two examples illustrate the distinctions between Milnor and asymptotically stable attractors. The first one exhibits a Milnor attractor which is not asymptotically stable, while the next one exhibits an asymptotically stable attractor which is not a minimal Milnor attractor.

Example 1.5.4 We consider the widely studied *Feigenbaum map*; see for instance Guckenheimer [39], Eckmann *et al.* [28] and Milnor [69]. This is the quadratic map $f(x) = \lambda x(1-x)$ of the unit interval $[0,1]$ at $\lambda = 3.57\ldots$, the accumulation of period-doubling bifurcations – often called the Feigenbaum point. This map has an invariant transitive (indeed topologically minimal) Cantor set K whose basin of attraction is the whole interval $[0,1]$ except for the denumerably many periodic orbits of period 2^n and their denumerably many preimages [39]. It follows that $\mathcal{B}(K)$ has full Lebesgue measure, and therefore K is a Milnor attractor; more than that, it is a Melbourne attractor (in fact, it has even stronger attraction properties than a Melbourne attractor as $\ell(\mathcal{B}(A) \cap V)/\ell(V) = 1$ for every neighbourhood V of K). K is also Liapunov stable.

However, this Cantor set K cannot be asymptotically stable, by Theorem 1.4.6. We also know by Remark 1.3.12 that K must be accumulated by ω-limit sets other than itself. The non-wandering set (see Shub [108]) is $K \cup \mathrm{Per}(f)$, see Milnor [69]; so for each $y \in K$ there must exist a sequence of periodic points $\{x_n\}_{n\geq 0}$ such that $x_n \to y$. Thus $\mathcal{B}(K)$ contains no neighbourhood of K. In fact, even more is true. As preimages of the periodic points are dense in $[0,1]$ (see Dellnitz *et al.* [25]), it follows that $\mathcal{B}(K)$ contains no open set. □

Example 1.5.5 Let $f: I \to I$ be a continuous map of the compact interval I such that I is topologically transitive under f and $K \subset I$ is a transitive 'fat' Cantor set under f – that is, K has positive one-dimensional Lebesgue measure. Such a map can be explicitly constructed, for example, by adapting Bowen's [16] construction of a transitive 'fat' Cantor set. Let $0 < r < 1$ and define $F: I \times [-1,1] \to I \times [-1,1]$ by

$$F(x,y) = (f(x), ry)\,.$$

Then I is an asymptotically stable attractor for F. I admits an invariant stable foliation whose leaves are the vertical lines $\{x\} \times [-1,1]$, with $x \in I$. Thus the basin of attraction of an arbitrary invariant set $B \subset A$ is $B \times [-1,1]$. As the Cantor set K has positive one-dimensional measure, it follows that its basin $\mathcal{B}(K)$ has positive two-dimensional measure. Thus I does not satisfy (2) in Definition 1.4.7 and therefore is not an attractor in the sense of Milnor. □

The following example illustrates Remark 1.4.11: it exhibits a minimal Milnor attractor which is not topologically transitive.

Example 1.5.6 The key idea is to consider a flow on the plane with a saddle point which has a homoclinic cycle. Inside this cycle there is a point repellor and no other ω-limit sets. All orbits starting inside the cycle other than the point repellor will have as ω-limit set the whole homoclinic connection, which is obviously not transitive. The following is a translation of this phenomenon into the language of discrete dynamics.

Let X be the closed unit disc D^2. Using polar coordinates, we define a homeomorphism $f : D^2 \to D^2$ by

$$f(r,\theta) = (\sqrt{r},\ 2\pi(\theta^2 + 1 - r) \mod 2\pi).$$

The only invariant sets are the origin, the boundary circle C and the fixed point $p = (1,0)$. All points in the interior of D^2 except $(0,0)$ spiral outwards under the action of f. The dynamics on the circle C has a fixed point $p = (1,0)$ and all other points in C move clockwise under f, their orbits converging to p. Thus for $x \in C$ we have $\omega(x) = p$. Consider a small open ball B_p in D^2 centered in p. It is easy to show that all points $x \in B_p \setminus C$ eventually iterate away from B_p. By compactness $\omega(x) \cap D^2 \setminus B_p$ is non-empty. But $\omega(x)$ is invariant by Proposition 1.2.4 and $\omega(x) \neq (0,0)$ as the origin is a repellor. This implies $\omega(x) = C$.

Thus $\omega(x) = C$ for an open set of D^2. However, for all $x \in C$, $\omega(x) = \{p\}$; moreover the basin of attraction of $\{p\}$ is C, which has zero two-dimensional measure. Therefore C is a minimal Milnor attractor which is not topologically transitive. □

We next present an example of a Milnor attractor which is not *Liapunov* stable (and therefore not asymptotically stable) but still is a Melbourne attractor. This is the example originally presented by Melbourne which motivates his definition [66]. This example is set in the context of flows of ODE's but may be turned into a discrete dynamical system by considering an adequate Poincaré section, along the lines of Example 1.5.6.

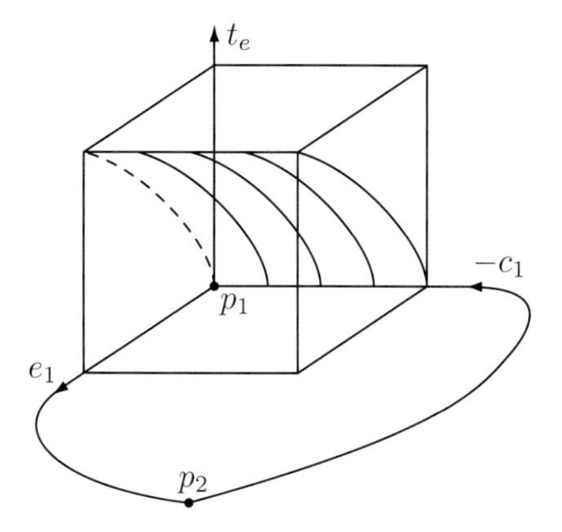

Figure 1.4: The cuspoidal region of instability and the region of asymptotic stability.

Example 1.5.7 Consider a flow giving rise to the invariant set shown in Figure 1.4. The set X is a heteroclinic cycle consisting of two equilibria and two connecting trajectories. Each equilibrium has a stable eigenvalue $-c_i$ and an unstable eigenvalue e_i in the directions along X.

Suppose the heteroclinic cycle is embedded in a flow on a 3-manifold. If the third eigenvalue at the fixed point have opposite signs, then under certain conditions (see Melbourne [66]) there exists a cuspoidal region C abutting the heteroclinic connections joining p_1 to p_2 with the following property. Given any neighbourhood V of the heteroclinic cycle, there is an open neighbourhood V such that trajectories starting in $V \setminus C$ remain in $U \setminus C$ in forward time and are asymptotic to the heteroclinic cycle. □

The next example is that of a Milnor attractor which has an open basin of attraction, therefore containing a neighbourhood of the attractor, but which is Liapunov unstable (and therefore is not an asymptotically stable attractor). This is the reason why we avoid the terminology "open-basin attractor" (see comments in § 1.6).

If we are to construct a Liapunov unstable set A with an open basin of attraction, this set A must possess some kind of homoclinic connection – in any neighbourhood, however small, of A there are points which wander away from it but eventually converge to it, as by the open-basin condition their ω-limit set is contained in A. There are many ways of constructing such an example; once more we choose the simplest one. Chapter 3 will in due course present more interesting and rich examples arising in a natural way; this example is the simplest exhibiting what we define in Chapter 3 to be an attractor with a *locally riddled basin*.

Example 1.5.8 Let $f : S^1 \to S^1$ be the circle map defined by its lift to the universal cover $\mathbb{R}$

$$F(x) = x + \frac{1}{2\pi}(1 - \cos(2\pi x)),$$

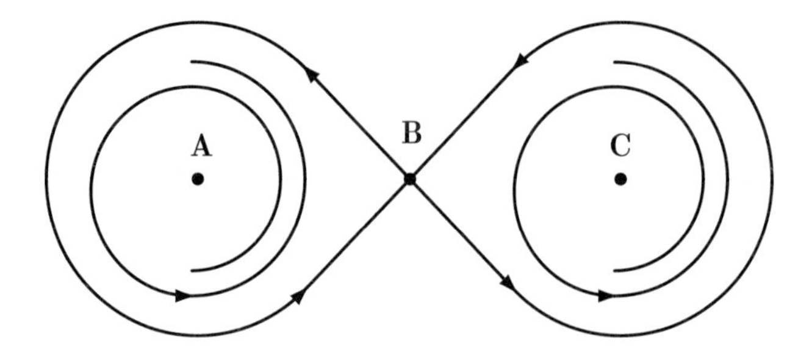

Figure 1.5: Bowen's counterexample: an SBR measure which is not strongly SBR.

where the usual projection $\Pi : \mathbb{R} \to S^1$, $\Pi(x) = e^{2\pi i x}$ is used. f has a unique fixed point $\theta = 0$ which is non-hyperbolic and one-sidedly stable. For all $\theta \in S^1$, $\omega(\theta) = \{1\}$ under f; thus $\mathcal{B}(\{1\}) = S^1$. On the other hand, $\{1\}$ is clearly Liapunov unstable since any $x \notin \mathbb{N}$ satisfies $F^n(x) \to [x] + 1$. □

Our last example is due to Bowen and illustrates the fact that strong SBR measures are a strictly stronger concept than SBR measures. It is set in the language of flows of differential equations but is straightforwardly translated into the language of maps by taking the time-1 map. Specific formulae may be derived, as in Example 1.5.6.

Example 1.5.9 Consider a flow in $\mathbb{R}^2$ with three fixed points A, B, C, where A and C are sources and B is a saddle with a double homoclinic connection (see fig 1.5). Then any initial condition except for the equilibria A, C gives rise to a time average which converges to δ_B. This measure is clearly *not* absolutely continuous along the unstable manifold W^u. □

1.6 Historical remarks and further comments

Remark 1.6.1 (Liapunov and Asymptotic Stability) The question of stability of motion in ordinary differential equations is an old and important problem. However, a general framework for the study of the stability of motion becomes possible only with the advent of the qualitative theory of ODE's in the late nineteenth century, mainly through the works of Poincaré [94] and Liapunov [62].

Poincaré's memoir, or rather series of memoirs, can be considered the starting point of the geometric theory of ODE's and modern dynamical systems theory. In most of his memoirs Poincaré limits himself to two-dimensional systems. His concept of stability is what could be called "orbital stability":

> Nous dirons que la trajectoire d'un point mobile est stable lorsque, décrivant autour du point de départ un cercle ou une sphère de rayon r, le point mobile, après être sorti de ce cercle ou de cette sphère, y

> rentrera une infinité de fois, et cela, quelque petit que soit r *(Jour. Math. Pur. App., 4.ème série, t. 1, pg. 171).*

Poincaré's *point mobile* is, of course, the position of the orbit in phase space. So orbital stability means in modern language that $x \in \omega(x)$.

Liapunov's seminal memoir *On the General Problem of the Stability of Motion* acknowledges his profound debt to the methods pioneered by Poincaré. Indeed, Liapunov states that

> ... when we know how to integrate our differential equations, this problem certainly presents no difficulty. But it will be important to have methods which permit it to be resolved independently of the possibility of this integration *(pg. 531).*

This memoir is set on the framework of ordinary differential equations. Liapunov starts out by pointing the inadequacy of previous studies on the subject, which essentially studied stability of a system by performing a linear stability analysis and proving on a case-by-case basis, if at all, that the full non-linear system would have the same stability properties. He then defines formally what he means by *stability of a solution of a differential equation*; this is what is known to us at present as Liapunov stability. In a further section he states and proves what we call today *Liapunov's Stability Theorem*, where Liapunov functions are first introduced. This Theorem states that, if a solution $x(t)$ preserves a positive definite function $V(x)$ and, for all y in a small enough neighbourhood of $x(t)$, the time derivative $\dot{V}(y) \leq 0$, then $x(t)$ is stable. Moreover, if $\dot{V}(y) < 0$ for $x \neq y$, then $x(t)$ is asymptotically stable. In actual fact Liapunov does not use the term "asymptotically stable" – his expression in the statement of the Theorem is that "the disturbed motion approaches the undisturbed one asymptotically".

Besides Liapunov and asymptotic stability, it is also in this work that Liapunov exponents are first introduced. Liapunov refers to them as "characteristic numbers" or "characteristic exponents".

Remark 1.6.2 (Attractors) As we have seen, the definition of attractor in a differentiable dynamical system is not an obvious matter. In our opinion there is no "correct" definition of attractor. Different definitions emphasize different properties; it is preferable to have a description of the behaviour of the attractor with respect to all these properties than to just one. For instance, we know much more about the Feigenbaum attractor by the description in 1.5.4 than just by knowing whether or not it satisfies a specific condition, as asymptotic stability or being a minimal Milnor attractor.

Our point of view is that, as long as the two "programmatic" conditions of § 1.4 are satisfied in some rigorously defined sense, the precise definition of "attractor" is largely a matter of mathematical purpose and convenience.

The generic "programmatic" conditions of § 1.4 seem to be well accepted. They were stated explicitly by Ruelle [103]. He first imposes three conditions – invariance, closedness or compactness and topological attractivity – in his definition. He agrees to call such sets "attracting sets" – as opposed to "attractors", which must satisfy an irreducibility condition as well. The nature of this irreducibility condition may vary; examples are topological transitivity or being the support of an ergodic measure (which of course implies topological transitivity by Proposition 1.3.3). Although, as we have shown, Ruelle's topological conditions are not necessarily satisfied in the measure-theoretic setting, they may be taken in the abstract as a 'programme' to be realized by specific definitions, working in specific categories. In this sense Milnor's definitions 1.4.7 and 1.4.8 fulfil Ruelle's general programme.

The review presented in this chapter is not meant to be exhaustive. Instead, it concentrates on the careful definition and characterization of the concepts of attractor which will be directly relevant to our further work. Several other meaningful and important definitions are possible; we briefly and broadly describe two of them.

- **Conley-Ruelle attractors [103].** Given a continuous function $f : X \to X$ on a compact metric space, write $a \succ b$ if, for any $\epsilon > 0$, there is an ϵ-pseudo-orbit going from a to b. Then the relation $a \succ b$ is a pre-order; writing $a \sim b$ if both $a \succ b$ and $b \succ a$ we obtain an equivalence relation. The relation $\succ$ induces an order relation on the equivalence classes of the latter, which we denote by $[a] \geq [b]$ if $a \succ b$. Then $[a]$ is a *basic class* if a is a fixed point or $\mathrm{card}[a] > 1$. The union of the equivalence classes is Conley's *chain recurrent attractor* [24]. A minimal equivalence class under the order $\geq$ is a *Conley-Ruelle attractor* [103].
  Several results may then be drawn for these Conley-Ruelle attractors; see [103]. The motivation for this definition is that, while all the others considered so far are based on orbits, in this one the attractor is constructed from *pseudo*-orbits. It may be argued that only pseudo-orbits are observed in either physical or numerical experiments, experimental and/or numerical round-off errors effectively imposing small random perturbations; see Ruelle [103].

- **Generic Milnor attractors [69].** A different definition of attractor was proposed by Milnor. This definition runs along similar lines as that in Definition 1.4.7 but relies only in topological concepts; essentially the concept of "largeness" is not measured by Lebesgue measure but by Baire category. More specifically, the definition is obtained from Definition 1.4.7 by replacing "set of positive Lebesgue measure" by "residual set". Milnor then explores the relationships between these (almost everywhere) limit sets in the two senses.

Remark 1.6.3 (Axiom A attractors) Axiom A attractors, and more generally Axiom A systems, were introduced by Smale [113] in what is probably the most famous paper on Dynamical Systems. As previously discussed, Axiom A attractors are asymptotically stable. In addition, the hyperbolic structure implies a number of essential properties for these systems: a canonical decomposition of the manifold in a finite number of basic sets satisfying the Spectral Decomposition Theorem; everywhere existence, within the basic sets, of local stable and unstable manifolds; the shadowing property (implying that Axiom A attractors are Conley-Ruelle attractors); existence of Markov partitions; existence of an SBR measure for Axiom A attractors.

These properties endow Axiom A attractors with an extremely rich dynamical structure. Much of the mathematical work in Dynamical Systems during the 1960's was, on the one hand, to explore the mathematical properties of Axiom A systems and, on the other, try to show that they constitute a 'large' class of dynamical systems. Ideally one would like to show that Axiom A systems were generic – that is, would include a residual set in $\mathrm{Diff}^{\,r}(M)$, thus encompassing 'almost all dynamical systems'. We would then have a complete understanding of 'almost all complex dynamics'.

This was essentially the programme outlined and pursued by the Smale school. However, it gradually became clear that it was impossible to realize, and that for an understanding of 'almost all complex dynamics' it is necessary to go beyond the hyperbolic theory.

The first step in this direction was provided by Smale himself. It is more or less clear that structural stability implies Axiom A (indeed, the proof of the Stability Conjecture by Mañé [64] in 1988, showing that structural stability is equivalent to Axiom A and strong transversality, proves that the two concepts are essentially equivalent). Smale [112] exhibited in 1966 an open set of structurally unstable diffeomorphisms of the 3-torus $\mathbb{T}^3$ – showing that structurally stable systems are not dense, therefore not residual, in $\mathrm{Diff}^{\,r}(M)$, where M is a compact manifold. The general problem of characterization of structural stability was therefore more subtle than the corresponding one for 2-dimensional flows, solved in 1962 by Peixoto [88].

During the 1970's the situation gradually became clearer. Several important studies showed that non-hyperbolic attractors can occur in a *persistent* way – that is, as residual subsets of open sets in $\mathrm{Diff}^{\,r}(M)$. These include for instance the studies of the geometric Lorenz attractor of Guckenheimer and Williams [42] and the results of Newhouse ([75], [76], [77]) on homoclinic bifurcations – that is, tangencies of stable and unstable manifolds – in hyperbolic systems, proving the existence in a persistent way of systems with infinitely many periodic attractors and of wild hyperbolic sets (hyperbolic sets in which these tangencies occur).

Apart from these developments, which were in a sense internal to hyperbolic theory, research in dynamics acquired for several reasons a different character during this period. The outcome was the systematic development of new techniques and methods aiming at the study of non-uniformly hyperbolic systems. Thus, for instance, the growing influence of Ergodic Theory in dynamics.

Remark 1.6.4 (Open-basin attractors) One of the problems plaguing the notion of attractor is the Babel of existing definitions. Different authors assign widely different meaning to expressions such as "attracting set", "attractor", "open-basin attractor". This is particularly true in the applications-oriented literature, where often terminology is not explicitly defined at the outset.

The term *open-basin attractor*, in particular, has frequently been used to refer to the objects we call asymptotically stable attractors. We avoid using it for reasons which are clear from our discussion: whereas it is true that asymptotically stable attractors have an open basin, the converse is not only false (recall example 1.5.8) as context-dependent. In particular, an open basin of attraction in principle implies nothing in terms of stability; asymptotic stability is a strictly stronger condition than openness of the basin of attraction. Even if we restrict the expression "open-basin attractor" to mean "asymptotically stable attractor", it is still misleading – it does not characterize accurately the object under consideration.

Remark 1.6.5 (Milnor attractors) Milnor's 1985 papers ([69], [70]) are the first references which emphasize in formal definitions the role of positive measure of the basin of attraction. More than solving problems, these papers raise them – and are definitely a landmark in the concept of attractor.

Although these definitions were motivated by known examples, it later became clear that Milnor's concept of attractor was the best suited to describe phenomena associated to dynamical systems arising in certain applications – in fields as varied as synchronization of chaotic oscillators, symmetric chaos or lattice coupled systems (see e.g. Pikovski *et al.* [91], Alexander *et al.* [1], Ashwin *et al.* [4], Buescu *et al.* [20], Melbourne [66]). Milnor attractors arise in a very natural way in certain contexts; they are indeed to be expected generically in certain circumstances (see chapter 3). They also show up very naturally in our Chapters 2 and 3.

We next present several independent comments about Milnor's papers.

Basin of attraction. Milnor originally proposed to rename the basin of attraction $\mathcal{B}(A) = \{x \in X : \omega(x) = A\}$ the *realm of attraction* in the unstable case, reserving the term basin of attraction for the case where $\mathcal{B}(A)$ is an open set. The purpose was to distinguish explicitly the measure-theoretical attractors, which may be unstable, from asymptotically stable attractors.

However, this term did not become standard, mainly because of the already loose usage of the term 'basin'. By 1993 Alexander *et al.* [1] acknowledged this fact, when introducing the phenomenon of *riddled basins:* working with unstable Milnor attractors, they refer to Milnor's papers and explicitly avoid the term *realm* – "else our paper would be titled 'Riddled Realms' ".

Irreducibility. Milnor's original setting is where X is a compact manifold. It is then easy to see from Definition 1.4.7 that every continuous map has at least one attractor. Other than that, an arbitrary union of disjoint attractors is still an attractor. This definition therefore does not incorporate any irreducibility in the sense of Ruelle. Consider for instance the identity map on a compact interval. Then any compact subinterval (indeed any compact subset of positive measure) is an attractor, as is the closure of an arbitrary union of these.

The definition of a minimal Milnor attractor incorporates an irreducibility requirement: no proper compact subset of A is an attractor. As pointed out in Remark 1.4.11 and Examples 1.5.5 and 1.5.6, this condition neither implies nor is implied by topological transitivity. There is no guarantee that minimal attractors exist at all, as shown by the identity map example. However, at most denumerably many minimal attractors may exist (because all the basins are disjoint, have positive measure and X is compact).

A cautionary note: the Bowen horseshoe. A Smale horseshoe is necessarily a saddle-like object. Keeping the discussion in two dimensions for clarity, it is a transitive Cantor set in which every point has a one-dimensional local stable manifold and a one-dimensional unstable manifold. As the unstable manifolds cannot be contained in the Cantor set, it is Liapunov unstable. In fact, the methods developed in the next chapter will show that an invariant Cantor set on which the dynamics is conjugate to a horseshoe cannot be embedded in a dynamically stable way in any finite-dimensional manifold – even in the absence of hyperbolicity and invariant manifolds. Thus a horseshoe can never be a topological attractor. In terms of the basic set decomposition, a horseshoe "can never lie at a vertex of the stability diagram" [113]. Bowen and Ruelle [18] prove that the basin of attraction of a basic set Ω for a C^2 Axiom A diffeomorphism has zero Lebesgue measure – unless the basic set is an (asymptotically stable) attractor. Thus in the C^2 case the only possible hyperbolic Milnor attractors are the Axiom A attractors.

Surprisingly, the C^2 condition *is* necessary – a C^1 horseshoe *can* be an attractor in Milnor's sense. Bowen [16] constructed a horseshoe for a C^1 diffeomorphism of the 2-disk in the standard way, with the only difference that he allows the Cantor set to be "thickened" – that is, at each step in the Cantor set construction he deletes a decreasing proportion of the remaining intervals. Controlling this rate of decrease allows the construction of a hyperbolic Cantor set of positive 1-dimensional measure which can be embedded in a diffeomorphism. The Cartesian product of this Cantor set with itself is a Cantor set of positive two-dimensional measure, whose basin consequently also has positive two-dimensional measure. Therefore this "Bowen horseshoe", although a topological saddle, is a Milnor attractor.

Remark 1.6.6 (Melbourne attractors) These are not a new type of attractors but, in fact, a subclass of Milnor attractors on which a stronger condition is imposed. We define them because they will appear naturally in our later work.

Remark 1.6.7 (Components of attractors) Dellnitz, Golubitsky and Melbourne [25] were the first to observe that analysis of connected components of invariant sets has remarkable consequences for dynamics. They prove a topological lemma which implies a number of interesting corollaries. The most remarkable general results thus obtained are:

1. If A is a topologically mixing stable ω-limit set, it is connected.
2. If A is a stable ω-limit set containing a point of period k, then A has at most k connected components.
3. If a stable ω-limit set contains a periodic point, it cannot be a Cantor set.

The conclusions may be strengthened in the case of one-dimensional dynamics: let $X = \mathbb{R}$ or S^1 and $f : X \to X$ be continuous. Then:

1. If $\omega(x)$ is topologically transitive, then $\omega(x) \subset \overline{\mathrm{Per}(f)}$.
2. If $\omega(x)$ is compact and not minimal, then $\omega(x)$ has weak dependence; if $\omega(x)$ has positive measure it has sensitive dependence.
3. If $\omega(x)$ consists of a finite union of closed intervals, then it is topologically transitive, periodic points are dense and there is sensitive dependence.

For a definition of sensitive dependence, see Guckenheimer [39], Dellnitz *et al.* [25] or § 2.6 below. For a definition of weak dependence, see Dellnitz *et al.* [25].

Hirsch [46] then proved a result which is essentially equivalent to our Theorem 1.4.6: asymptotically stable attractors have finitely many connected components which are cyclically permuted. Furthermore, by a suitable modification of the standing assumptions (replacing local compactness of X by compactness of the operator $f : X \to X$) the theorem may be adapted to the case where X is a Banach space. This result may then be applied to a semilinear second order parabolic equation with space variables in a smooth compact manifold M, smooth boundary conditions and periodic data. Assume there is a Banach space E of functions on M and a semiflow $S = \{S_t\}_{t \geq 0}$ on $E \times Q$, where β is the period, Q denotes the circle $\mathbb{R}/\beta\mathbb{Z}$, and each S_t is a compact nonlinear operator. The map f is the restriction of S_β to $X = E \times p$, for any $p \in Q$.

Assuming S has a compact asymptotically stable ω-limit set $L \subset E \times Q$ under f, and setting $A = L \cap X$, it follows that A has a finite number ν of connected components. Moreover, if L contains an orbit of period $m\beta$ for some positive integer m, then ν divides m.

Remark 1.6.8 (SBR attractors) Our definition of SBR measure amounts to what is referred to sometimes as the *natural* (see Rand *et al.* [96]) or *physical* (see Eckmann *et al.* [28]) measure. The term SBR, or Sinai-Bowen-Ruelle, is sometimes reserved for what we call strong SBR measures, although there is no standard convention. For instance, our choice coincides with those of Young [121] or of Gambaudo *et al.* [33].

The exceptional property of SBR measures is that, starting with a random point (with respect to Lebesgue measure) in a small enough neighbourhood of A, the time averages will converge to the phase average given by μ *with positive probability*. It is also easy to see (Newhouse [78]) that, if B is an open set with

$\mu_{SBR}(B) = \mu_{SBR}(\overline{B})$, then

$$\lim_{n\to\infty} \frac{1}{n} \sum_{i=0}^{n-1} \chi_B(f^j(x)) = \mu_{SBR}(B)$$

for a.e. x in the basin of A. This means that we can approximate phase averages by the box-counting procedure.

The construction of SBR measures is an extremely difficult problem. The instances in which they are known to exist require uniform or at least non-uniform hyperbolicity. They can be explicitly constructed for Axiom A attractors (see Ruelle [100]) and for expanding maps of the interval (Lasota and Yorke [60]) or of the circle (Shub and Sullivan [109]), where there is some form of uniform hyperbolicity. Under some conditions in the quadratic family of maps of the interval, non-uniform hyperbolicity of subsets of the non-wandering set may be achieved (Misiurewicz [73], Jakobson [51], Benedicks and Carleson [11]), thereby allowing the construction of strong SBR measures. In the two-dimensional case Young [121] has constructed SBR measures for Lozi-type maps, and recently Benedicks and Young [13] have done the same for the Hénon map, under less stringent non-uniformly hyperbolic conditions.

A common feature to all known cases is that the only way to prove existence of an SBR measure is by proving it is strongly SBR. However, there is the conviction that they should not be an exceptional occurrence even in the non-hyperbolic case. Although somewhat speculative, this notion is supported by numerical evidence (see Dellnitz *et al.* [26]), generally by box-counting.

Chapter 2
Liapunov Stability and Adding Machines

2.1 Introduction

In Chapter 1 we discussed several notions of stability for compact invariant sets of dynamical systems. Here we shall prove that, under very general hypotheses, the set of connected components of a stable set of a discrete dynamical system possesses a tightly constrained structure. More precisely, suppose that X is a locally compact, locally connected metric space, $f : X \to X$ is a continuous mapping (not necessarily invertible) and A is a compact transitive set. Let K be the set of connected components of A and let $\tilde{f} : K \to K$ be the map induced by f. We proved in § 1.3 that either K is finite or a Cantor set; in either case $\tilde{f}$ acts transitively on K. Our main result (Theorem 2.3.1 below) is that, if A is Liapunov stable and has infinitely many connected components, then $\tilde{f}$ acts on K as a 'generalized adding machine', which we describe in a moment. We remark that imposing the stronger condition of asymptotic stability destroys the Cantor structure altogether and K must be finite – which is the content of Theorem 1.4.6. Thus adding machines can be Liapunov stable but *never* asymptotically stable. This Theorem may be strengthened to a version that does not require transitivity but the weaker property of being a stable ω-limit set.

This part of our work originated in a paper of Dellnitz *et al.* [25] on the structure of attractors (indeed stable ω-limit sets) with finitely many components. They prove, among other results, that $\tilde{f}$ permutes these components cyclically. Here we analyse the case where A is a stable set with infinitely many components. An important source of motivation is the remark by Milnor [69] that the dynamics of the quadratic map $x \mapsto \lambda x(1-x)$ at the Feigenbaum point is a binary adding machine. Adding machines, which are defined via symbolic dynamics, are a class of minimal maps on Cantor sets.

We consider *generalized adding machines*, defined thus. Let $\underline{k} = (k_n)_{n>0}$ be a sequence of integers $k_n \geq 2$. Let $\Sigma_{\underline{k}}$ be the set of all one-sided infinite sequences $\underline{i} = (i_n)_{n>0}$ for which $0 \leq i_n < k_n$. Think of these sequences as 'integers' in multibase notation, the base of the nth digit i_n being k_n. With the natural (product) topology, $\Sigma_{\underline{k}}$ is homeomorphic to the Cantor set. We define a map $\alpha_{\underline{k}} : \Sigma_{\underline{k}} \to \Sigma_{\underline{k}}$, which informally may be described as 'add 1 and carry' where the addition is performed at the leftmost term i_1 and the carry proceeds to the right in multibase notation. It is well known that $\alpha_{\underline{k}}$ is a minimal homeomorphism; we provide a proof in § 2.2.

We now state the main results of this chapter. Let X be a locally compact locally connected metric space, $f : X \to X$ a continuous mapping. Suppose that A is a compact transitive set with infinitely many connected components. Assume that A is Liapunov stable. As in § 1.3, define an equivalence relation $\sim$ on A by setting $x \sim y$ if and only if x and y lie in the same connected component of A, and let $K = A/\sim$ be the quotient space with the identification topology. Because f maps connected sets to connected sets, it induces a continuous map $\tilde{f} : K \to K$. We know by Theorem 1.3.14 that K is a Cantor set on which $\tilde{f}$ acts transitively. We shall prove a much stronger result under the assumption of stability: there exists a sequence $\underline{k}$ such that the dynamics on K is topologically conjugate to that of $\alpha_{\underline{k}}$ on $\Sigma_{\underline{k}}$. The key step is in § 2.3, where we prove the existence of non-trivial $\tilde{f}$-invariant finite partitions of K. In §2.5 we show that the result is still true if A is supposed to be only a stable ω-limit set, not necessarily transitive. The dynamics on K is conjugate to those on a suitable inverse limit space of such partitions.

We give a complete classification of adding machines up to topological conjugacy using spectral methods as in Walters [118]; other proofs with a more combinatorial or topological flavour may be constructed. We show that every generalized adding machine occurs for some map f defined on the closed disk $D \subset \mathbb{R}^2$ (or equivalently the 2-sphere S^2).

This chapter is organized as follows. The basic facts on adding machines, their symbolic dynamics, and ergodic theory are recalled in § 2.2. The main theorem is proved in § 2.3 in the transitive case, with two basic ingredients: the fact that stability of A implies minimality of $\tilde{f}$, and the existence of the aforementioned finite invariant partitions of K. In §2.4 we show the implications of stable adding machines for the set of periodic points in the special case of the real line. In §2.5 we show in which sense adding machines are inverse limit maps on inverse limit spaces, and we take advantage of this construction in § 2.6 to extend the main theorem to the case of stable ω-limit sets. We classify generalized adding machines in § 2.7 by studying the eigenvalues of $\tilde{f}$, which form a countable subgroup of the unit circle. We prove that this subgroup is a complete invariant, and deduce a simpler combinatorial invariant based on the multiplicity (finite or infinite) with which a given prime occurs as a factor in the sequence $\underline{k}$. In § 2.8 we construct, for any $\underline{k}$, a map of the closed disk $D \subset \mathbb{R}^2$ for which the dynamics of $\tilde{f}$ on K is topologically conjugate to that of $\alpha_{\underline{k}}$ on $\Sigma_{\underline{k}}$.

2.2 Adding Machines and Denjoy maps

In this section we recall the abstract definition, via symbolic dynamics, of the class of maps of the Cantor set known as *adding machines.*

Let X be a topological space. We shall say that a subset $S \subset X$ is *clopen* if it is simultaneously closed and open in X.

Let $\underline{k} = \{k_n\}_{n\geq 1}$ be a sequence of integers with $k_n > 1$ for all n. Let

$$\Sigma_{\underline{k}} = \Pi_{n=1}^{\infty}\{0, \dots, k_n - 1\}$$

be the space of one-sided infinite sequences $\underline{i} = (i_n)_{n\geq 1}$ such that $0 \leq i_n < k_n$ equipped with the product topology. It is easy to verify that $\Sigma_{\underline{k}}$ is metrizable; indeed, a metric compatible with this topology is $d(\underline{i}, \underline{j}) = \sum_{n=1}^{\infty} |i_n - j_n|/k_n^n$. An elementary check shows that $\Sigma_{\underline{k}}$ is homeomorphic to the Cantor set.

We define the *adding machine* with *base* $\underline{k} = (k_1, k_2, \dots)$, which we denote by $\alpha_{\underline{k}} : \Sigma_{\underline{k}} \to \Sigma_{\underline{k}}$, as follows. Let $(i_1, i_2, \dots) \in \Sigma_{\underline{k}}$. Then

$$\alpha_{\underline{k}}(i_1, i_2, \dots) = \begin{cases} (\underbrace{0, \dots, 0}_{l-1}, i_l + 1, i_{l+1}, \dots) & \text{if } \; i_l < k_l - 1 \text{ and } i_j = k_j - 1 \text{ for } j < l \\ (0, 0, \dots, 0, \dots) & \text{if } \; i_j = k_j - 1 \text{ for all } j. \end{cases}$$

It is well known that $\alpha_{\underline{k}}$ is a minimal homeomorphism of $\Sigma_{\underline{k}}$. Indeed, continuity and bijectivity are immediate, and since $\Sigma_{\underline{k}}$ is compact and Hausdorff, $\alpha_{\underline{k}}$ is a homeomorphism. To establish minimality, take arbitrary $j > 0$, $\underline{a}, \underline{b} \in \Sigma_{\underline{k}}$. For $i = 1, \dots, j$, let $n_i = a_i - b_i \pmod{k_i}$, and let $N_j = n_1 + \sum_{i=2}^{j} k_{i-1} n_i$. Then $\alpha_{\underline{k}}^{N_j}(\underline{a})$ and $\underline{b}$ agree in their first j entries, and therefore $\alpha_{\underline{k}}^{N_j}(\underline{a}) \to \underline{b}$ as $j \to \infty$. Thus $\underline{b} \in \omega(\underline{a}, \alpha_{\underline{k}})$ and $\alpha_{\underline{k}}$ is minimal.

It is also interesting to consider adding machines from the ergodic point of view. Define the *cylinder sets of length* n on $\Sigma_{\underline{k}}$ by

$$C_{n;i_1,\dots,i_n} = \{\underline{x} \in \Sigma_{\underline{k}} : x_1 = i_1, \dots, x_n = i_n\}.$$

Cylinder sets are clopen and have the obvious properties that

$$\alpha_{\underline{k}}(C_{n;i_1,\dots,i_n}) = \begin{cases} C_{n;0,\dots,0,i_l+1,i_{l+1},\dots,i_n} & \text{if } \; i_j = k_j - 1 \text{ for } j < l \text{ and } i_l = k_l - 1 \\ C_{n;0,0,\dots,0} & \text{if } \; i_j = k_j - 1 \text{ for } j = 1, \dots, n, \end{cases} \tag{2.1}$$

and therefore

$$\alpha_{\underline{k}}^{k_1 \dots k_n}(C_{n;i_1,\dots,i_n}) = C_{n;i_1,\dots,i_n}, \tag{2.2}$$

that is, the $k_1 \dots k_n$ cylinders of length n form a finite invariant partition of $\Sigma_{\underline{k}}$ by clopen sets.

The set of all cylinders generates the topology on $\Sigma_{\underline{k}}$, and therefore the Borel σ-algebra $\mathcal{B}$. Equation (2.1) and the fact that $\alpha_{\underline{k}}$ is a homeomorphism together imply that any $\alpha_{\underline{k}}$-invariant measure μ must assign equal measure to all cylinders of length n, and therefore

$$\mu(C_{n;i_1,\dots,i_n}) = \frac{1}{k_1 \dots k_n}.$$

By standard approximation arguments (see e.g. Mañé [65]), knowledge of μ restricted to all cylinders determines a unique $\mu \in C(\Sigma_{\underline{k}})^*$. Therefore $\Sigma_{\underline{k}}$ has a unique Borel $\alpha_{\underline{k}}$-invariant measure. In other words $(\Sigma_{\underline{k}}, \alpha_{\underline{k}}, \mathcal{B})$ is uniquely ergodic. If $(\Sigma_{\underline{k}}, \alpha_{\underline{k}})$ is regarded as a topological group in the natural way (see § 2.5), then the unique invariant measure constructed above is exactly Haar measure by unique ergodicity.

We shall also be interested in another well known class of minimal homeomorphisms of the Cantor set, arising from the so-called *Denjoy maps* of the circle. We sketch their construction; a more rigorous account may be found in Schweitzer [106].

Let $S^1 = \mathbb{R}/\mathbb{Z}$ with the usual metric, and let $R_\omega(x) = x + \omega \pmod 1$ be rigid rotation by $\omega \in \mathbb{R} \setminus \mathbb{Q}$. Take any $x_0 \in S^1$ and, for $n \in \mathbb{Z}$, replace $R_\omega^n(x_0)$ by a small interval I_n, such that $\sum_{n \in \mathbb{Z}} \ell(I_n) < \infty$, where ℓ denotes Lebesgue measure. Map I_n onto I_{n+1} by an orientation-preserving homeomorphism. The resulting map is a homeomorphism of S^1, which we denote by D_ω, such that $\bigcup_{n \in \mathbb{Z}} \operatorname{int} I_n$ is a dense collection of open disjoint intervals, which are easily seen to be wandering. Its complement K is an invariant Cantor set such that $\omega(x, D_\omega) = K$ for all $x \in K$, or indeed in S^1, see Arrowsmith and Place [3]. So K is minimal under D_ω. It is possible to choose the I_n and corresponding homeomorphisms so that D_ω is a C^1 diffeomorphism (see Schweitzer [106]), but not C^2 (see Guckenheimer and Holmes [40]).

This construction seems somewhat artificial, but Denjoy maps occur naturally in smooth, even analytic, dynamical systems – for instance, in the breakup of invariant topological circles of irrational rotation number into cantori in Hamiltonian dynamics (see for instance MacKay and Meiss [63]).

The rotation number ω is a complete invariant for Denjoy maps, in the sense that two Denjoy maps D_{ω_1}, D_{ω_2} are topologically conjugate if and only if $\omega_1 = \omega_2$. Necessity is implied by the fact that the rotation number is a topological invariant; sufficiency is seen from their very construction – just map corresponding I_n homeomorphically and extend this by continuity to a homeomorphism of S^1, which is a conjugacy.

Denjoy maps are *not*, however, topologically conjugate to any adding machine map. This is a direct corollary of the results proved in § 2.7, from which another proof of the sufficiency assertion in the preceding paragraph will be derived.

2.3 Stable Cantor sets are Adding Machines

We now state and prove the main result of this chapter. Recall the construction and notations of § 1.3: suppose a continuous map f of a (locally connected, locally compact) metric space X admits a compact transitive set A. Then we quotient A by its connected components and define $K = A/\sim$ to be the quotient space and $\tilde{f} : K \to K$ to be the map induced on the quotient space by f.

Theorem 2.3.1 *Suppose that X is a locally connected, locally compact metric space, $f : X \to X$ is a continuous map, and A is a compact transitive set. Assume A is Liapunov stable and has infinitely many components. Then $\tilde{f} : K \to K$ is topologically conjugate to an adding machine.*

It is convenient to prove first minimality of $\tilde{f}$ and then use this to prove Theorem 2.3.1. Minimality is proved in Lemma 2.3.3. We isolate a purely topological result used in this proof in the proposition below.

Proposition 2.3.2 *Let $(K, \tilde{d})$ be the metric Cantor set, and F a closed non-empty proper subset of K. Then there exists $C \supset F$ with C clopen and $C^c \neq \emptyset$. That is, $C \cup C^c$ form a disconnection of K with $F \subset C$.*

Proof. We use the following standard result from point-set topology: a compact Hausdorff space K is totally disconnected if and only if it has an open base whose sets are also closed; see Simmons [110].

Because F is closed and is a proper subset of K, which is also closed, there exist $\epsilon > 0$ and $z \in K \setminus F$ such that $\tilde{d}(z, F) \geq \epsilon$. Take $G = \bigcup_{x \in F} G_r(x)$, where the $G_r(x)$ are clopen sets containing x such that $G_r(x) \subset B_{\epsilon/3}(x)$. The set G is open, and the $\{G_r(x)\}$ form an open cover of F. As F is closed it is also compact, hence there exists a finite subcover $\{G_{r_1}(x_1), \dots, G_{r_n}(x_n)\}$. Set $C = \bigcup_{j=1}^{n} G_j$. Then C is clopen because all the C_j are, and $F \subset C$. On the other hand, by construction $z \notin G_r(x)$ for all x, hence $z \in C^c$, so $C^c \neq \emptyset$. Therefore C, C^c have all the required properties. □

The next lemma proves the minimality assertion of Theorem 2.3.1.

Lemma 2.3.3 *Under the conditions of Theorem 2.3.1, $\tilde{f} : K \to K$ is minimal.*

Proof. Suppose for a contradiction that K is not minimal under $\tilde{f}$; that is, there exists $z \in K$ such that $\omega(z, \tilde{f}) = \tilde{F} \neq K$. Then $\tilde{F}$ is closed in K, or equivalently $F = \pi^{-1}(\tilde{F})$ is closed in A; therefore $F \neq A$. Construct a disconnection $C \cup C^c$ of K as in Lemma 2.3.2; that is, choose C so that both C, C^c are non-empty, clopen and $\tilde{F} \subset C$. Define $C_1 = \pi^{-1}(C), C_2 = \pi^{-1}(C^c)$; it follows that C_1, C_2 are disjoint, non-empty and both (relatively) clopen in A. Moreover, $\pi^{-1}(\tilde{F}) \subset \pi^{-1}(C) = C_1$.

Since A is closed in X, it follows that C_1, C_2 are closed and disjoint in X. Because X is a metric space it is normal, so there exist disjoint open sets $U_1' \supset C_1$, $U_2' \supset C_2$. Then $U' = U_1' \cup U_2'$ is an open neighbourhood of A with $F \subset U_1'$.

By Propositions 1.3.5 and 1.3.6 and Remark 1.3.7 there is an open neighbourhood U of A, $U \subset U'$, $U = U_1 \cup \ldots \cup U_k \cup U_{k+1} \cup \ldots \cup U_n$ with $\bigcup_{j=1}^{k} U_j \subset U_1'$ and $\bigcup_{j=k+1}^{n} U_j \subset U_2'$, and all U_j being open, disjoint, connected, non-empty and with disjoint compact closure. Thus $F \subset C_1 \subset \bigcup_{j=1}^{k} U_j$.

Take an open $V \subset U$ as in the definition of Liapunov stability; that is, for all $n > 0$, $f^n(V) \subset U$. Again by Propositions 1.3.5 and 1.3.6, by eventually shrinking V we may suppose it to be of the form $V = \bigcup_{i=1}^{l} V_i$, with the V_i open, connected, non-empty and pairwise disjoint. As V is a neighbourhood of A, there is at least one V_i such that $V_i \cap F \neq \emptyset$; by renumbering if necessary, suppose it is V_1.

Connectedness of the V_i, U_j, continuity of f and the stability condition imply that, for all $n > 0$, $f^n(V_i)$ is contained inside a single U_j. By density of transitive points, there is a transitive point $x \in V_1$, and thus a subsequence $\{n_k\}$ such that $f^{n_k}(x) \in U_{k+1}$, whence by continuity $f^{n_k}(V_1) \subset U_{k+1}$. But $V_1 \cap F \neq \emptyset$, so taking $y \in V_1 \cap F$ it follows that $f^{n_k}(y) \in U_{k+1}$. Therefore, there exists a subsequence $f^{n_{k_j}}(y)$ converging to some $z \in \overline{U_{k+1}}$. Thus $z \in \omega(y)$ but $z \notin \bigcup_{j=1}^{k} U_j$ and so $z \notin F$, contradicting compactness and invariance of F. □

Proof of Theorem 2.3.1. Let U be a neighbourhood of A with compact closure consisting of a finite number of disjoint connected open sets, all of them intersecting A, whereas U itself is not connected, and such that the Hausdorff distance between connected components of A lying inside the same component of U is smaller than 1 (since A is compact, only a finite number of its components may be a distance greater or equal than 1 apart). That is, $U = \bigcup_{j=0}^{m} U_j$, where $m \geq 1$ and the U_j are disjoint, open, connected, intersect A and, if A_1, A_2 are components of A lying in the same U_j, then $d_H(A_1, A_2) < 1$. By stability, we may choose an open neighbourhood $V \subset U$ of A such that $f^n(V) \subset U$ for all $n > 0$, in such a way that $V = \bigcup_{i=0}^{l} V_i$, where the V_i are open, connected, disjoint and have non-empty intersection with A. Because $f^n(V_i)$ is connected and contained in U, it follows that for all $i = 0, \ldots, l$ there is a unique $j(n,i)$, $j = 0, \ldots m$, such that $f^n(V_i) \subset U_{j(n,i)}$.

Consider the projection π acting on $U_i \cap A, V_j \cap A$. Defining $\tilde{U_i} = \pi(U_i \cap A)$, $\tilde{V_j} = \pi(V_j \cap A)$, and recalling that Haudorff distance between components projects via π to the distance $\tilde{d}$ in K (see Lemma 1.3.13), these statements translate to the following conditions on K:

$$K = \bigcup_{i=0}^{m} \tilde{U_i} = \bigcup_{j=0}^{l} \tilde{V_j}, \quad \tilde{U_i},\ \tilde{V_j} \text{ clopen in } K; \tag{2.3}$$

$$\forall n > 0\ \tilde{f}^n(\tilde{V_i}) \subset \tilde{U}_{j(n,i)}; \tag{2.4}$$

$$\text{if } x_1, x_2 \in \tilde{U_j} \text{ then } \tilde{d}(x_1, x_2) < 1. \tag{2.5}$$

For convenience of notation, we henceforth omit the tildes, keeping in mind that the U_i, V_j now represent clopen sets in K.

Since each V_j is contained in a single U_i, condition (2.3) may be rewritten as $U_i = V_{i_1} \cup \ldots \cup V_{i_{l_i}}$, with obviously $\sum_{i=0}^{m} l_i = l$.

We now construct an invariant partition of K from the V_i.

Suppose that for a certain i_k we have $\tilde{f}(V_{i_k}) \subset U_j$, and there are distinct j_1, j_2 such that $\tilde{f}(V_{i_k}) \cap V_{j_1} \neq \emptyset$, $\tilde{f}(V_{i_k}) \cap V_{j_2} \neq \emptyset$. By construction $\tilde{f}^n(V_{i_k})$ is contained in a single $U_{j(n,i_k)}$ for all $n \geq 0$, so this forces all of $\tilde{f}^{n+1}(V_{i_k})$, $\tilde{f}^n(V_{j_1})$, $\tilde{f}^n(V_{j_2})$ to be contained in a single $U_{j(n+1,i_k)}$ for all $n \geq 0$. Now define $W_{1;j_1} = V_{j_1} \cup V_{j_2}$; this set still retains the properties of being clopen in K and $\tilde{f}^n(W_{1;j_1})$ being contained in a single $U_{j(j_1,n)}$ for all $n \geq 0$.

Perform this procedure for all V_i, and note that there are only a finite number of these and they are all clopen. We arrive, eventually after renumbering the $W_{1;j}$, at a finite partition $\mathcal{W} = \{W_{1;0}, \ldots, W_{1;k_1-1}\}$ of K, with $W_{1;i}$ clopen and $\bigcup_{i=0}^{k_1-1} W_{1;i} = K$, such that

$$\forall n \geq 0,\ 0 \leq i < k_1,\ \tilde{f}^n(W_{1;i}) \subset W_{j(n;i)}, \tag{2.6}$$

because of condition (2.4) and the facts that each U_i is the union of finitely many $W_{1;j}$ and the image of a given $W_{1;j}$ intersects a single $W_{1;k}$. Note also that each $W_{1;j}$ is contained inside a single U_i and is a subset of K. Condition (2.5) then implies that each $W_{1;j}$ has diameter smaller than 1 (the diameter of W is simply $\sup_{x_1,x_2 \in W} \tilde{d}(x_1, x_2)$.)

Minimality of $\tilde{f}$ implies that condition (2.6) is actually an equality, as otherwise there would exist an invariant proper non-empty closed subset of K. This implies that $\tilde{f}(W_{1;i}) = W_{1;j(i)}$, for some $j(i)$. But, again by minimality, this function $j : \{0, \ldots, k_1 - 1\} \to \{0, \ldots, k_1 - 1\}$ must be a cyclic permutation. We can therefore renumber the sets $W_{1;i}$ in such a way that the following condition holds:

$$\tilde{f}(W_{1;i}) = W_{1;i+1} \pmod{k_1}. \tag{2.7}$$

Notice that each $W_{1;i}$, $i = 0, \ldots, k_1 - 1$, is a Cantor set, is invariant and minimal under $\tilde{f}^{k_1}$, and $\pi^{-1}(W_{1;i})$ is Liapunov stable under f^{k_1} – that is, satisfies the same properties relative to f^{k_1} as K itself relative to f. This means we can play exactly the same game at a new level, and refine the partition $\mathcal{W}$ to a new partition which preserves these properties.
To be specific, set $W_{1;0} = W_{2;0} \cup \ldots \cup W_{2,k_2-1}$, with $W_{2;j}$ clopen and $\tilde{f}^{k_1}(W_{2;j}) = W_{2;j+1} \pmod{k_2}$. Now $\tilde{f}(W_{1;0}) = W_{1;1}$, so that

$$\tilde{f}(W_{2;0}) \cup \ldots \cup \tilde{f}(W_{2;k_2-1}) = W_{1;1}.$$

Indeed, by minimality and as a result of our construction of the $W_{1;i}$, the $\tilde{f}(W_{2;j})$ are disjoint, so they partition $W_{1;1}$. We show below this partition to be clopen.

For $0 \leq i_1 < k_1$, $0 \leq i_2 < k_2$, define $W_{2;i_1,i_2} = \tilde{f}^{i_1 k_1}(W_{2;i_2})$. This partition of K is easily seen to have the following properties: first, $W_{2;i_1,i_2} \subset W_{1;i_1}$ for all

$0 \le i_2 < k_2$ and $\bigcup_{i_2=0}^{k_2-1} W_{2;i_1,i_2} = W_{1;i_1}$; second, the action of $\tilde{f}$ on the elements of this partition is

$$\tilde{f}(W_{2;i_1,i_2}) = \begin{cases} W_{2;i_1+1,i_2} & \text{if } \ i_1 < k_1 - 1 \\ W_{2;0,i_2+1 \pmod{k_2}} & \text{if } \ i_1 = k_1 - 1. \end{cases}$$

In particular, $\tilde{f}^{k_1 k_2}(W_{2;i_1,i_2}) = W_{2;i_1,i_2}$ for all $i_1 < k_1$, $i_2 < k_2$. The fact that every $W_{2;i_1,i_2}$ is clopen now follows from the facts that $W_{2;0,i_2}$ is clopen, satisfies this last relation and $\tilde{f}$ is continuous. As above, compactness of K and the fact that this partition is finite implies that we can perform all these choices subject to the condition that the diameter of each $W_{2;i_1,i_2}$ is smaller than $1/2$.

Proceeding inductively, we construct a family of partitions of K into clopen sets $W_{n;i_1,\dots,i_n}$ with the properties:

$$W_{n;i_1,\dots,i_n} = \bigcup_{i_{n+1}=0}^{k_{n+1}-1} W_{n+1;i_1,\dots,i_{n+1}}, \tag{2.8}$$

the $W_{n+1;i_1,\dots,i_{n+1}}$ being disjoint, having diameter less than $1/n$ and satisfying

$$\tilde{f}(W_{n;i_1,\dots,i_n}) = \begin{cases} W_{n;0,\dots,0,i_l+1,i_{l+1},\dots,i_n} & \text{if } \ i_j = k_j - 1, \ j < l \text{ and } i_l < k_l - 1 \\ W_{n;0,0,\dots,0} & \text{if } \ i_j = k_j - 1 \text{ for } j = 1,\dots,n. \end{cases} \tag{2.9}$$

Given any sequence $\{i_j\}_{j\ge 1}$ with $0 \le i_j < k_j$, consider $H = \bigcap_{j=1}^{\infty} W_{j;i_1,\dots,i_j}$. As this is an intersection of nested compact sets it must be non-empty by the finite intersection property. Furthermore, our construction ensures that the diameter of H is zero. Hence H is a singleton.

Define, as in the previous section, $\Sigma_{\underline{k}}$ as $\Pi_{j=1}^{\infty}\{0,\dots,k_j - 1\}$ with the product topology, and $h : K \to \Sigma_{\underline{k}}$ by $h(x) = (i_1, i_2, \dots)$, where the sequence $\{i_j\}$ is that defined by $\{x\} = \bigcap_{j=1}^{\infty} W_{j;i_1,\dots,i_j}$. The argument above shows that h is a bijection. To show it is continuous, note that an open ball around an element

$$(i_1, i_2, \dots) = \underline{i} \in \Sigma_{\underline{k}}$$

is the set of sequences $\underline{i}'$ such that $i'_1 = i_1, \dots, i'_n = i_n$ for some integer n; its inverse image under h is $\bigcap_{j=1}^{n} W_{j;i_1,\dots,i_j}$, which is open in K.

Thus, h is a continuous bijection from a compact space onto a Hausdorff space, and therefore is a homeomorphism.

It remains to show that h is a topological conjugacy between $\tilde{f}$ and $\alpha_{\underline{k}}$, or in other words that the diagram

$$\begin{array}{ccc} K & \xrightarrow{\tilde{f}} & K \\ h\downarrow & & \downarrow h \\ \Sigma_{\underline{k}} & \xrightarrow{\alpha_{\underline{k}}} & \Sigma_{\underline{k}} \end{array}$$

commutes. To do so, suppose that $x \in K$ is such that $i_l(h(x)) < k_l - 1$ and $i_j(h(x)) = k_j - 1$ for all $j < l$. Then $\tilde{f}(x) = \bigcap_{j=1}^{\infty} W_{j;0,\dots,0,i_l+1,i_{l+1},\dots,i_j}$ by equation (2.9), and thus $i_j(h(\tilde{f}(x))) = 0$ for $j < l$, $i_l(h(\tilde{f}(x))) = i_l(h(x)) + 1$, and, for all $j > l$, $i_j(h(\tilde{f}(x))) = i_j(h(x))$. If $i_j(h(x)) = k_j$ for all j, it follows from (2.9) that $\{\tilde{f}(x)\} = \bigcap_{j=1}^{\infty} W_{j;0,0\dots,0,\dots}$, and therefore $i_j(h(\tilde{f}(x))) = 0$ for all $j \geq 1$. Then, by definition, $\alpha_{\underline{k}} = h \circ \tilde{f} \circ h^{-1} : \Sigma_{\underline{k}} \to \Sigma_{\underline{k}}$ is the adding machine map on $\Sigma_{\underline{k}}$. □

Remark 2.3.4 We can extract a useful general principle from the proof of Theorem 2.3.1. Suppose that a map g on a metric space Y possesses a sequence of finite invariant partitions $\mathcal{P}_j$, $j \geq 1$, such that $\mathcal{P}_{j+1}$ refines $\mathcal{P}_j$ for all j. Let $p_j = |\mathcal{P}_j|$. Suppose that:

1. g acts transitively on each $\mathcal{P}_j$ (hence cycles its elements);
2. $p_{j+1} \geq 2p_j$ for all j;
3. If $P_j \in \mathcal{P}_j$ for all $j \geq 1$ and $P_1 \supseteq P_2 \dots$, then $\bigcap_{j=1}^{\infty} P_j$ is a singleton.

Then the method of proof of Theorem 2.3.1 shows that g is conjugate to the adding machine $\alpha_{\underline{k}}$, where $k_1 = p_1$ and $k_j = p_j/p_{j-1}$ for $j \geq 2$.

We use this remark in §2.8. □

We can immediately derive a corollary that generalizes a theorem of Hirsch [46], the difference being that we deal with *stable transitive* sets rather than attractors. Our corollaries apply, for instance, to the logistic map at the Feigenbaum point. Another generalization of Hirsch's result may be found in Dellnitz *et al.* [25].

Corollary 2.3.5 *Let A be a compact stable transitive set under f and suppose that f admits a periodic point of period n. Then A has k connected components, where k divides n.*

Proof. If x is periodic for f, then $\pi(x)$ is periodic for $\tilde{f}$. By Lemma 2.3.3 K is finite, say $K = \{1, \dots, k\}$, and $\tilde{f}$ must be a cyclic permutation by Theorem 1.3.14. Hence $\pi(x)$ is periodic with period k, and k divides n. □

Corollary 2.3.6 *Let A be a compact transitive set with infinitely many components. If A has a periodic point, then A is Liapunov unstable.*

Proof. Immediate from 2.3.5. □

In particular any compact invariant set on which $\tilde{f}$ is topologically conjugate to some symbolic subshift must violate the mild condition of Liapunov stability.

Corollary 2.3.7 *If A is a transitive stable Cantor set, then $f_{|A}$ is conjugate to an adding machine. In particular, $f_{|A}$ is a homeomorphism.*

Proof. Take $A \equiv K$, $f \equiv \tilde{f}$ in Theorem 2.3.1. □

In particular, we completely classify stable transitive sets for maps of the interval or of the real line:

Corollary 2.3.8 *Let A be a compact stable transitive set for a continuous map $f : \mathbb{R} \to \mathbb{R}$ or $f : I \to I$, I a compact interval. Then either A is a periodic orbit, or a finite cycle of intervals, or a Cantor set on which f is conjugate to an adding machine.* □

Corollary 2.3.9 *The minimal Cantor set K_ω for any Denjoy map D_ω is Liapunov unstable.*

Proof. This follows from the fact that, for all $\omega \in \mathbb{R} \setminus \mathbb{Q}$, D_ω is not topologically conjugate to any adding machine map, as will be shown in §2.7. □

Corollary 2.3.10 *Let A be a stable transitive set with infinitely many components. Then for all $x \in A$, $\omega(x, f)$ intersects all components of A.*

Proof. If not, then $\omega(\pi(x), \tilde{f})$ is a non-empty compact invariant proper subset of K, contradicting minimality. □

If A is itself minimal, as happens when A is a Cantor set, then Corollary 2.3.10 is vacuous. We can, however, construct examples where A is not minimal – for instance, as follows.

Example 2.3.11 Let $I = [0, 1]$, $F : I \to I$ be the logistic map at the Feigenbaum point, let $\varphi : I \to I$ be given by $\varphi(x) = 4x(1 - x)$, and consider the product map $f : I \times I \to I \times I$ for which $f(x, y) = (F(x), \varphi(y))$. Then f has a stable invariant set of the form $A = K \times I$, where K is the stable invariant Cantor set associated to the first factor. It is easy to show that $f_{|K \times I}$ is transitive. Indeed, as F is conjugate to the binary adding machine and φ to a full 2-shift, it is easy to locate transitive points explicitly via symbolic dynamics.

The components of A are the sets of the form $\{\alpha\} \times I$, for $\alpha \in K$. It is also easy to check that A is Liapunov stable. However, 0 is a fixed point for $\varphi : I \to I$, and therefore $K \times \{0\}$ is a closed invariant subset of $K \times I$, which is nowhere dense in $K \times I$ and intersects all components of $K \times I$. More generally, periodic points of φ form a dense F_σ in I (density of the set of periodic points is well known; this set is also denumerable. Alternatively, its complement contains the transitive points, which form a dense G_δ by Lemma 1.3.1). To each periodic orbit $\{x, \varphi(x), \ldots, \varphi^{n-1}(x)\}$ of φ there corresponds a closed invariant nowhere dense subset $K \times \{x, g(x), \ldots, g^{n-1}(x)\}$ of $K \times I$, which as required intersects all components of A. Thus the invariant transitive subsets of A form in this case a dense F_σ in A.

2.4 Adding Machines and periodic points: interval maps

The purpose of the next sections is to study several questions related to the existence of stable adding machines. The first of these is the implication of a stable adding machine in terms of periodic points. The second is the conjugacy problem for adding machines: as it is clear from the proof of Theorem 2.3.1, there is a certain degree of freedom when performing the choices of partitions leading to the base sequence $\underline{k}$ characterizing the adding machine. Performing different choices will lead to adding machines with a different base sequence $\underline{k}'$, say. However, it follows from the statement of the Theorem itself that the dynamical systems $(\Sigma_{\underline{k}}, \alpha_{\underline{k}})$ and $(\Sigma_{\underline{k}'}, \alpha_{\underline{k}'})$ are topologically conjugate. This fact leads naturally to the problem of characterizing conjugacy classes of adding machines.

This section is devoted to the first problem in a particular setting: that of maps of the interval. For a discussion of periodic points and stable adding machines in the general case, see Remark 2.9.6.

We first need some terminology. Let X be, as above, a locally connected locally compact metric space and $f : X \to X$ be a continuous map. Suppose that f admits a stable transitive set on which the quotient map $\tilde{f}$ acts as an adding machine. We call the sequence of U, V as constructed in § 2.3 *tame neighbourhoods* of A. Tame neighbourhoods have the crucial property of projecting to K under the identification map as a clopen finite invariant partition whose elements are cycled according to (2.1). We call such a sequence of nested tame neighbourhoods an *invariant nest*.

An invariant nest defines a *tame sequence* $\underline{k} = (k_1, k_2, \dots)$. The order of the elements of an invariant nest is thus $k_1, k_1k_2, \dots, \Pi_{j=1}^n k_j, \dots$.

We have the following result about periodic points:

Theorem 2.4.1 *Let $f : I \to I$ be a continuous map of the interval admitting a transitive stable Cantor set K on which f is conjugate to an adding machine $\alpha_{\underline{k}}$, where $\underline{k}$ is a tame sequence. Let $x \in K$. Then there is a sequence $x_n \in I \setminus K$ with $x_n \to x$ such that x_n is periodic with period $p_n = \Pi_{j=1}^n k_j$. In particular $K \subset \overline{Per(f)}$.*

Proof. Let $h : K \to \Sigma_{\underline{k}}$ be the homeomorphism establishing conjugacy. Let $C_{1;i_1}$ be the length 1 cylinder containing $h(x)$, $K_{1;i_1} = h^{-1}(C_{1;i_1})$ be the corresponding clopen subset of K and $H_{1;i_1}$ be the closed convex hull of $K_{1;i_1}$ in I, that is, the smallest closed interval containing $K_{1;i_1}$. Following the proof of Theorem 2.3.1, we see that

$$f^n(H_{1;i_1}) \supseteq H_{1;i_1+n \pmod{k_1}}, \tag{2.10}$$

while Liapunov stability implies that

$$f^n(H_{1;i_1}) \cap H_{1;i_1} = \emptyset, \quad 1 \le n \le k_1 - 1. \tag{2.11}$$

As the endpoints of $H_{1;i_1}$ belong to $K_{1;i_1}$ which is invariant under f^{k_1} and $H_{1;i_1}$ is connected, it follows that $f^{k_1}(H_{1;i_1}) \supseteq H_{1;i_1}$. By the fried-egg lemma below, f^{k_1} must possess a fixed point x_1 in $H_{1;i_1}$, which must be a point of period k_1 by the cycling condition (2.11).

For the inductive step, suppose we have located periodic points $x_1, \dots, x_n$ of periods $k_1, \dots, \Pi_{j=1}^{n} k_j$ in a neighbourhood of x. The closed convex hull of the cylinder of length $n+1$ around x, which we denote by $H_{n+1;i_1,\dots,i_{n+1}}$, satisfies equations similar to (2.10) and (2.11) with k_1 replaced by $\Pi_{j=0}^{n+1} k_j$. So there is a fixed point x_{n+1} for $f^{k_1 \dots k_{n+1}}$ in $H_{n+1;i_1,\dots,i_{n+1}}$. The cycling condition (2.11) again ensures x_{n+1} is not fixed by any previous iterate of f, and therefore x_{n+1} has prime period $\Pi_{j=1}^{n+1} k_j$. As $\bigcap_{n=1}^{\infty} H_{n;i_1,\dots,i_n} = \{x\}$, it follows that $x_n \to x$, as asserted. □

Remark 2.4.2 Application of Theorem 2.4.1 to the Feigenbaum map shows not only that F admits all periods that are powers of 2 but the stronger fact that any point on the invariant Cantor set is the limit point of a sequence of periodic points with period 2^n.

The location of periodic points requires in general the use of a suitable fixed-point theorem, as periodic points of period n are fixed by the nth iterate of the map. In the proof of Theorem 2.4.1 this is supplied by the fried-egg lemma below.

Lemma 2.4.3 (The real fried-egg) *Let $f : \mathbb{R} \to \mathbb{R}$ be a continuous map. Suppose I is a compact interval such that $f(I) \supseteq I$. Then f has a fixed point in I.*

Proof. If not, then for all $x \in I$ either $f(x) > x$ or $f(x) < x$. In the first case continuity and compactness imply there exists $\epsilon > 0$ such that, for all $x \in I$, $f(x) > x + \epsilon$. It follows that $f(I)$ cannot cover I, contradicting the assumption. The second case is identical. □

We call the covering property $f(I) \supseteq I$ the *fried-egg property*. It is specific to $\mathbb{R}$ that the fried-egg property for continuous maps on compact balls implies existence of fixed points; counterexamples may be given in $\mathbb{R}^2$ – see Figure 2.1. Here Q is the unit square, $f(Q) \supseteq Q$ but f obviously has no fixed points. It is a simple matter to modify this example in such a way that $f(Q)$ is topologically the same as Q (say, by adding a twist which covers the hole). It is also clear that f may be chosen so that Q is Liapunov stable. Therefore Theorem 2.4.1 does not generalize to other spaces with the present methods.

Remark 2.4.4 An important point to keep in mind when searching for a generalization of Theorem 2.4.1 is that no general statement about periodic points is possible if the fibres over the adding machine – i. e. the components of A – are non-trivial. We give two examples to illustrate the situation; in the first one $\overline{\mathrm{Per}(f)} = \emptyset$, in the second one $A \subset \overline{\mathrm{Per}(f)}$. For simplicity we work with the Feigenbaum map of the interval $I = [0, 1]$ to which corresponds the binary adding machine; that

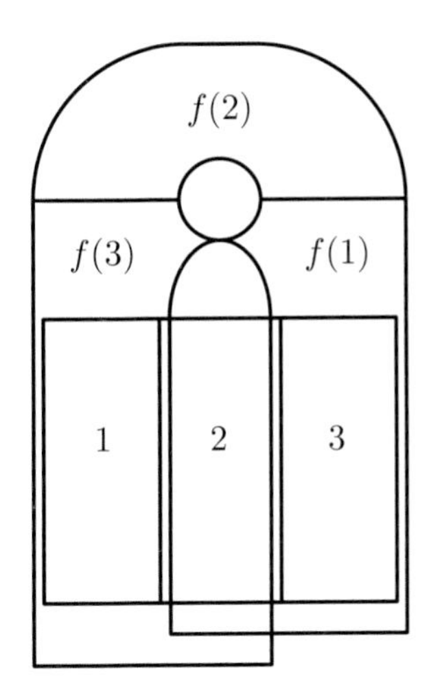

Figure 2.1: The unit square Q is divided into 3 vertical strips which map in the indicated way under f.

the corresponding results are valid for *arbitrary* adding machines is easily seen by replacing the Feigenbaum map by the example in § 2.8. □

Example 2.4.5 Let $f : I \times S^1 \to I \times S^1$ be defined by $f(x, y) = (F(x), R_\omega(y))$ where $F : I \to I$ is the Feigenbaum map and $R_\omega(y) = y + \omega \pmod 1$ is an irrational rotation. Let K be the invariant Cantor set for F; it is easily shown that $K \times S^1$ is a stable transitive set under f (sketch: stability is trivial. For transitivity, remark that every open set in $K \times S^1$ must contain a cylinder cross an interval in S^1. Then for every open U, V there is $n \geq 0$ such that $f^n(U) \cap V \neq \emptyset$, which is equivalent to topological transitivity). The only other ω-limit sets under f, whose existence is implied by Remark 1.3.12, are those of the form

$$A_n = \{x_n, F(x_n), \ldots, F^{2^n-1}(x_n)\} \times S^1,$$

x_n being the unique period 2^n orbit of F. As $\omega \in \mathbb{R} \setminus \mathbb{Q}$ it is clear that $f_{|A_n}$ is minimal; in particular f has no periodic points. □

Example 2.4.6 Let $f : I \times I \to I \times I$ be defined by $f(x, y) = (F(x), \varphi(y))$, where again F is the Feigenbaum map and $\varphi(y) = 4y(1 - y)$. Here $K \times I$ is a stable transitive set and is accumulated by invariant sets of the form $A_n = \{x_n, F(x_n), \ldots, F^{2^n-1}(x_n)\} \times I$. As periodic points are dense in each A_n, it follows that $K \times I$ lies in the closure of the periodic points. □

2.5 Interlude: Adding Machines as inverse limits

In this section we present another description of adding machines. As before, let $\underline{k} = (k_1, k_2, \ldots)$ be a sequence of integers with $k_j \geq 2$ for all j, and let $\alpha_{\underline{k}} : \Sigma_{\underline{k}} \to \Sigma_{\underline{k}}$ be the adding machine with base $\underline{k}$ as defined in § 2.2. Let

$$\begin{array}{rcl} p_1 & = & k_1 \\ p_2 & = & k_1 k_2 \\ \vdots & & \vdots \\ p_n & = & \Pi_{j=1}^{n} k_j \\ \vdots & & \vdots \end{array} \tag{2.12}$$

and let $(\mathcal{Z}_n, \overset{p_n}{\oplus})$ be the cyclic group of order p_n,

$$\mathcal{Z}_n \cong \mathbb{Z}_{p_n}.$$

We shall view $(\mathcal{Z}_n, \overset{p_n}{\oplus})$ as a topological group by endowing it with the discrete topology, and shall use the natural isomorphism with $\mathbb{Z}_{p_n}$ to use additive notation. Denote by $1_n \cong 1 \pmod{p_n}$ the natural generator of $(\mathcal{Z}_n, \overset{p_n}{\oplus})$.

There is a natural topological group homomorphism $\gamma_n : \mathcal{Z}_{n+1} \to \mathcal{Z}_n$ for $n \geq 1$, given by

$$\gamma_n(z) = z \pmod{p_n}.$$

This homomorphism is continuous, k_{n+1} to 1 and $\gamma_n(1_{n+1}) = 1_n$.

The sequence $\{\mathcal{Z}_n, \gamma_n\}$ thus defined is an inverse limit sequence of topological groups, whose inverse limit space we denote by $\mathcal{Z}_\infty^{\underline{k}}$. The following facts about $\mathcal{Z}_\infty^{\underline{k}}$ are immediately derived from the standard construction of inverse limit spaces, see e.g. Hocking and Young [48].

$\mathcal{Z}_\infty^{\underline{k}}$ is a topological group with respect to the induced topology and group operation. The induced topology is the subspace topology by considering it as a subspace of $\Pi_{j=1}^{\infty} \mathcal{Z}_j$ endowed with the Tychonoff topology. Relative to this topology $\mathcal{Z}_\infty^{\underline{k}}$ is compact, Hausdorff and non-empty.

The group operation – which we denote by $\oplus$ – between two elements of $\mathcal{Z}_\infty^{\underline{k}}$, say $\underline{i} = (i_1, i_2, \ldots)$ and $\underline{j} = (j_1, j_2, \ldots)$, is given by

$$\underline{i} \oplus \underline{j} = (i_1 \overset{p_1}{\oplus} j_1, i_2 \overset{p_2}{\oplus} j_2, \ldots),$$

relative to which $(\mathcal{Z}_\infty^{\underline{k}}, \oplus)$ is clearly abelian.

Given the element $\underline{1} = (1_1, 1_2, \ldots, 1_n, \ldots)$ of $\mathcal{Z}_\infty^{\underline{k}}$, we define a map

$$\mathcal{F} : \mathcal{Z}_\infty^{\underline{k}} \to \mathcal{Z}_\infty^{\underline{k}}$$

as translation by $\underline{1}$:

$$\mathcal{F}(z) = z \oplus \underline{1}.$$

Thus $(\mathcal{Z}_\infty^{\underline{k}}, \mathcal{F})$ defines a topological dynamical system. The main result in this section is that this dynamical system is topologically conjugate to the adding machine of base $\underline{k}$.

Proposition 2.5.1 *$(\mathcal{Z}_\infty^{\underline{k}}, \mathcal{F})$ is topologically conjugate to $(\Sigma_{\underline{k}}, \alpha_{\underline{k}})$.*

Proof. Let $\mathcal{S}_n = \{(i_1, \dots, i_n)\}$ be the set of all strings of symbols of length n, each i_j being an integer between 0 and $k_j - 1$. Endow each $\mathcal{S}_n$ with the discrete topology and define continuous maps $g_n : \mathcal{S}_{n+1} \to \mathcal{S}_n$ for $n \geq 1$ by

$$g_n(i_1, i_2, \dots, i_n, i_{n+1}) = (i_1, i_2, \dots, i_{n-1}, i_n).$$

Then the sequence $\{\mathcal{S}_n, g_n\}$ defines an inverse limit space $\mathcal{S}_\infty$ which is compact, Hausdorff and non-empty.

Define maps $\phi_n : \mathcal{S}_n \to \mathcal{Z}_n$ for $n \geq 1$ by

$$\phi_n\left((i_1, i_2, \dots, i_n)\right) = i_1 + p_1 i_2 + \dots + p_{n-1} i_n \pmod{p_n}.$$

Clearly the ϕ_n are 1-1, onto and continuous with a continuous inverse, hence homeomorphisms. Moreover the diagram

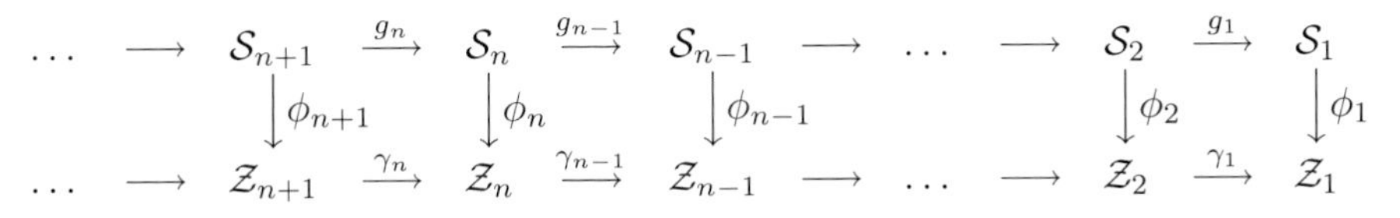

$$\begin{array}{ccccccccccccc}
\cdots & \longrightarrow & \mathcal{S}_{n+1} & \xrightarrow{g_n} & \mathcal{S}_n & \xrightarrow{g_{n-1}} & \mathcal{S}_{n-1} & \longrightarrow & \cdots & \longrightarrow & \mathcal{S}_2 & \xrightarrow{g_1} & \mathcal{S}_1 \\
 & & \downarrow \phi_{n+1} & & \downarrow \phi_n & & \downarrow \phi_{n-1} & & & & \downarrow \phi_2 & & \downarrow \phi_1 \\
\cdots & \longrightarrow & \mathcal{Z}_{n+1} & \xrightarrow{\gamma_n} & \mathcal{Z}_n & \xrightarrow{\gamma_{n-1}} & \mathcal{Z}_{n-1} & \longrightarrow & \cdots & \longrightarrow & \mathcal{Z}_2 & \xrightarrow{\gamma_1} & \mathcal{Z}_1
\end{array}$$

commutes. The $\{\phi_n\}$ thus induce a homeomorphism $\Phi : \mathcal{S}_\infty \to \mathcal{Z}_\infty^{\underline{k}}$ (both $\mathcal{S}_\infty$ and $\mathcal{Z}_\infty^{\underline{k}}$ are compact Hausdorff and Φ is clearly a continuous bijection).

The maps $\alpha_n : \mathcal{S}_n \to \mathcal{S}_n$ defined by

$$\alpha_n\left((i_1, \dots, i_n)\right) = \begin{cases} (0, 0, \dots, i_l + 1, i_{l+1}, \dots, i_n) & \text{if} \quad i_l < k_l - 1 \text{ and } i_j = k_j - 1 \text{ for } j < l \\ (0, 0, \dots, 0) & \text{if} \quad i_j = k_j - 1 \text{ for } j = 1, \dots, n, \end{cases}$$

similarly induce a mapping $\alpha_\infty : \mathcal{S}_\infty \to \mathcal{S}_\infty$: given $s \in \mathcal{S}_\infty$, $s = (i_1, i_2, \dots)$, then

$$\alpha_\infty(s) = \left(\alpha_1\left((i_1)\right), \alpha_2\left((i_1, i_2)\right), \dots\right).$$

A simple computation then shows that

$$\begin{aligned} \Phi \circ \alpha_\infty(s) &= (i_1 \overset{p_1}{\oplus} 1, i_1 \overset{p_2}{\oplus} p_1 i_2 \overset{p_2}{\oplus} 1, \dots) \\ &= \mathcal{F} \circ \Phi(s). \end{aligned} \tag{2.13}$$

Thus Φ is a topological conjugacy between $\alpha_\infty : \mathcal{S}_\infty \to \mathcal{S}_\infty$ and $\mathcal{F} : \mathcal{Z}_\infty^{\underline{k}} \to \mathcal{Z}_\infty^{\underline{k}}$.

We now sketch the construction of a homeomorphism conjugating α_∞ with the base $\underline{k}$ adding machine $\alpha_{\underline{k}}$. Define $\mathcal{C}_n$ to be the set of all cylinders of length n in $\Sigma_{\underline{k}}$. Define maps $h_n : \mathcal{S}_n \to \mathcal{C}_n$ by

$$h_n((i_1, \dots, i_n)) = C_{n; i_1, \dots, i_n}.$$

As above the h_n induce a homeomorphism H between the inverse limit spaces $\mathcal{S}_\infty$ and $\mathcal{C}_\infty$. However, the latter has a natural identification with $\Sigma_{\underline{k}}$. Moreover the equality

$$H(\alpha_\infty(s)) = \alpha_{\underline{k}}(H(s))$$

is seen by direct computation to hold. Thus $H^{-1} \circ \Phi$ is a homeomorphism conjugating $\alpha_{\underline{k}} : \Sigma_{\underline{k}} \to \Sigma_{\underline{k}}$ and $\mathcal{F} : \mathcal{Z}_\infty^{\underline{k}} \to \mathcal{Z}_\infty^{\underline{k}}$. □

2.6 Stable ω-limit sets are Adding Machines

In this section we use the construction performed in § 2.5 to prove a result which is slightly stronger than Theorem 2.3.1. Although the main statement is the same, the hypotheses are weaker: we only require stable ω-limit sets, which do not have to be *a priori* transitive.

Theorem 2.6.1 *Let X be a locally connected, locally compact metric space and $f : X \to X$ a continuous map. Suppose A is a stable ω-limit set under f with infinitely many components. Then the space of components K is a Cantor set on which the induced map $\tilde{f}$ is conjugate to an adding machine.*

Proof. By Liapunov stability (Lemma 1.3.8) and Propositions 1.3.5, 1.3.6 and Remark 1.3.7, there exists a basis of open neighbourhoods $\{U_\beta\}$ of A such that

1. U_β has compact closure,
2. $f(\overline{U}_\beta) \subset \overline{U}_\beta$,
3. $\overline{U}_\beta = \overline{U}_0 \cup \ldots \cup \overline{U}_{k_1-1}$, where the $\overline{U}_j$ are connected, disjoint and have non-empty intersection with A.

Fix one such U. Since A is an ω-limit set there exists $x \in U$ such that $\omega(x) = A$. Since $f(\overline{U}) \subset \overline{U}$ and the components $\overline{U}_j$ of $\overline{U}$ are connected and disjoint, the image $f(\overline{U}_i)$ of each $\overline{U}_i$ is contained in one and only one $\overline{U}_j$. As $A = \omega(x)$ intersects all $\overline{U}_j$, this implies that f acts on the set $\{\overline{U}_0, \ldots, \overline{U}_{k_1-1}\}$ as a cyclic permutation. We may relabel the $\overline{U}_j$ in such a way that

$$f(\overline{U}_i) \subset \overline{U}_{i+1} \pmod{k_1}.$$

We have thus constructed a compact neighbourhood of A which provides a covering $\mathcal{U}_1 = \{\overline{U}_{1;0}, \overline{U}_{1;2}, \ldots, \overline{U}_{1;k_1-1}\}$ of A such that

$$f(\overline{U}_{1;i}) \subset \overline{U}_{1;i+1} \pmod{k_1}. \tag{2.14}$$

However, by hypothesis A has infinitely many components, so at least one of the $\overline{U}_{1;j}$ must contain more than one component of A (in fact infinitely many by the pigeonhole principle). Suppose without loss of generality that this happens with $\overline{U}_{1;0}$. Since $A \cap \overline{U}_{1;0}$ is a stable ω-limit set under f^{k_1} we can apply the above construction to find a compact neighbourhood $\overline{U}_2 = \bigcup_{i_2=0}^{k_2-1} \overline{U}_{2;0,i_2}$ of $A \cap \overline{U}_{1;0}$ with $k_2 > 1$ connected components labeled by i_2, $0 \le i_2 \le k_2 - 1$ such that

$$f^{k_1}(\overline{U}_{2;0,i_2}) \subset \overline{U}_{2;0,i_2+1} \pmod{k_2}.$$

Furthermore, we may choose k_2 in such a way that the Hausdorff distance between any two components of $A \cap \overline{U}_{1;0}$ lying in the same connected component of U_2 is smaller than $1/2$ (by compactness, only finitely many components may be more than $1/2$ apart).

Defining $\overline{U}_{2;j,i_2} = f^j(\overline{U}_{2;0,i_2})$ for $1 \le j \le k_1 - 1$ yields a compact neighbourhood of A whose connected components satisfy

$$f(\overline{U}_{2;i_1,i_2}) \subset \begin{cases} \overline{U}_{2;i_1+1,i_2} & \text{if } i_1 < k_1 - 1 \\ \overline{U}_{2;0,i_2+1} \pmod{k_2} & \text{if } i_1 = k_1 - 1. \end{cases} \tag{2.15}$$

As components of A map among themselves in a 1-1 way, this argument incidentally shows that *all* $\overline{U}_{1;i_1}$ contain infinitely many components of A, thus showing there was no loss of generality in our assumption.

By applying the argument above to each $\overline{U}_{2;j,i_2}$ and refining we may suppose that the Hausdorff distance between any two components of A lying inside the same component of $\overline{U}_2$ is smaller than $1/2$.

Define $W_{2;i_1,i_2} = A \cap \overline{U}_{2;i_1,i_2}$. This is a clopen non-empty set in A (equipped with the relative topology). These sets are cyclically permuted by f according to

$$f(W_{2;i_1,i_2}) = \begin{cases} W_{2;i_1+1,i_2} & \text{if } i_1 < k_1 - 1 \\ W_{2;0,i_2+1} \pmod{k_2} & \text{if } i_1 = k_1 - 1. \end{cases} \tag{2.16}$$

Performing the quotient by connected components of A yields a finite clopen partition of $K = A/\sim$, invariant under $\tilde{f}$, whose elements $\tilde{W}_{2;i_1,i_2}$ are cycled according to (2.16).

We now proceed inductively to find a sequence $\{\overline{U}_n\}$ of compact neighbourhoods of A with the following properties:

1. $\overline{U}_n$ is forward-invariant;
2. the connected components of $\overline{U}_n$ are $\overline{U}_{n;i_1,\dots,i_n}$, each of which having non-empty intersection with A;
3. $\overline{U}_{n;i_1,\dots,i_n} \subset \overline{U}_{n-1;i_1,\dots,i_{n-1}}$;
4. the $\overline{U}_{n;i_1,\dots,i_n}$ are cycled according to

$$f(\overline{U}_{n;i_1,\dots,i_n}) \subset \begin{cases} \overline{U}_{n;0,0,\dots,i_l+1,i_{l+1},\dots,i_n} & \text{if } i_l < k_l - 1 \text{ and } i_j = k_j - 1 \text{ for } j < l \\ \overline{U}_{n;0,0,\dots,0} & \text{if } i_j = k_j - 1 \text{ for } j = 1,\dots,n; \end{cases} \tag{2.17}$$

5. the Hausdorff distance between any two components of A lying inside the same $\overline{U}_{n;i_1,\dots,i_n}$ is smaller than $1/n$.

As above, define $W_{n;i_1,\dots,i_n} = A \cap \overline{U}_{n;i_1,\dots,i_n}$. These are clopen, non-empty and cycled by f according to

$$f(W_{n;i_1,\dots,i_n}) = \begin{cases} W_{n;0,0,\dots,i_l+1,i_{l+1},\dots,i_n} & \text{if } i_l < k_l - 1 \text{ and } i_j = k_j - 1 \text{ for } j < l \\ W_{n;0,0,\dots,0} & \text{if } i_j = k_j - 1 \text{ for } j = 1,\dots,n. \end{cases} \tag{2.18}$$

Quotienting by connected components of A yields a finite clopen partition of $K = A/\sim$, invariant under $\tilde{f}$, whose elements $\tilde{W}_{n;i_1,\dots,i_n}$ are cycled according to (2.18).

Dropping the tildes by convenience, denote by $\mathcal{W}_n$ the cover of K given by the $\{W_{n;i_1,i_2,\dots,i_n}\}$. For each $n \geq 1$ define a continuous map $w_n : \mathcal{W}_{n+1} \to \mathcal{W}_n$ by

$$w_n(W_{n+1;i_1,i_2,\dots,i_{n+1}}) = W_{n;i_1,\dots,i_n},$$

which maps each element of $\mathcal{W}_{n+1}$ to the (unique) element of $\mathcal{W}_n$ containing it (as sets in K).

Thus $\{\mathcal{W}_n, w_n\}$ defines an inverse limit sequence whose inverse limit space we denote by $\mathcal{W}_\infty$. With the notation of § 2.5, defining maps $\psi_n : \mathcal{W}_n \to \mathcal{S}_n$ by

$$\psi_n(W_{n;i_1,\dots,i_n}) = (i_1,\dots,i_n)$$

induces a homeomorphism $\Psi : \mathcal{W}_\infty \to \mathcal{S}_\infty$ such that the diagram

$$\begin{array}{ccccccccccc} \cdots \longrightarrow & \mathcal{W}_{n+1} & \xrightarrow{w_n} & \mathcal{W}_n & \xrightarrow{w_{n-1}} & \mathcal{W}_{n-1} & \longrightarrow \cdots \longrightarrow & \mathcal{W}_2 & \xrightarrow{w_1} & \mathcal{W}_1 \\ & \downarrow \psi_{n+1} & & \downarrow \psi_n & & \downarrow \psi_{n-1} & & \downarrow \psi_2 & & \downarrow \psi_1 \\ \cdots \longrightarrow & \mathcal{S}_{n+1} & \xrightarrow{g_n} & \mathcal{S}_n & \xrightarrow{g_{n-1}} & \mathcal{S}_{n-1} & \longrightarrow \cdots \longrightarrow & \mathcal{S}_2 & \xrightarrow{g_1} & \mathcal{S}_1 \end{array}$$

commutes. This homeomorphism conjugates the induced map $\tilde{f}_\infty : \mathcal{W}_\infty \to \mathcal{W}_\infty$ to $\alpha_\infty : \mathcal{S}_\infty \to \mathcal{S}_\infty$. Thus by Proposition 2.5.1 the former is also topologically conjugate to the base $\underline{k}$ adding machine $\alpha_{\underline{k}} : \Sigma_{\underline{k}} \to \Sigma_{\underline{k}}$.

We next define a map $h : \mathcal{W}_\infty \to K$. If $p = (W_{1;i_1}, W_{2;i_1,i_2}, \dots)$ is a point in $\mathcal{W}_\infty$, the sets $W_{1;i_1}, W_{2;i_1,i_2}, \dots$ form a decreasing sequence of nested compact sets in K and thus have non-empty intersection. Our construction of the $W_{n;i_1,\dots,i_n}$ shows that the distance between any two points in $\bigcap_{n=1}^\infty W_{n;i_1,\dots,i_n}$, which by Lemma 1.3.13 is the Hausdorff distance between two different components of A in this intersection, is smaller than $1/n$ for all n. Therefore

$$\bigcap_{n=1}^{\infty} W_{n;i_1,\dots,i_n} = \{w\}$$

for a unique $w \in K$. Set $h(p) = w$.

First of all, h is injective. Indeed, if two points p and p' in $\mathcal{W}_\infty$ in the $n.^{th}$ coordinate, then $h(p) \neq h(p')$ because the elements of $\mathcal{W}_n$ are disjoint as sets in

K. Secondly, h is onto, since every $\mathcal{W}_n$ is a cover of K. Finally, h is continuous: the $W_{n;i_1,\dots,i_n}$ are clopen in K, form a basis for the topology and their inverse images under h are clopen in $\mathcal{W}_\infty$.

Thus h, being a continuous bijection from a compact to a Hausdorff space, is a homeomorphism. Again a simple calculation shows that $\tilde{f}_\infty : \mathcal{W}_\infty \to \mathcal{W}_\infty$ is topologically conjugate through h to $\tilde{f} : K \to K$, completing the proof. □

Remark 2.6.2 Although the statement of Theorem 2.3.1 is of course a particular case of Theorem 2.6.1, we should stress that, knowing only 2.3.1 and its proof, we would have no good reason to infer that it would generalize to ω-limit sets. Indeed ω-limit sets do not even have to be transitive, as our Example 1.5.2 shows. In fact, the statement of Theorem 2.6.1 might be interpreted as saying that Liapunov stability forces an ω-limit set with infinitely many components to be transitive.

We can easily adapt our Corollaries of Theorem 2.3.1 to this more general situation by replacing the requirement of A being a transitive set by that of being an ω-limit set.

Corollary 2.6.3 *Let A be a compact stable ω-limit set under f and suppose that f admits a periodic point of period n. Then A has k connected components, where k divides n.* □

Corollary 2.6.4 *Let A be a compact ω-limit set with infinitely many components. If A has a periodic point, then A is Liapunov unstable.* □

Corollary 2.6.5 *If A is a stable totally disconnected ω-limit set with infinitely many components, then A is a Cantor set and $f_{|A}$ is conjugate to an adding machine. In particular, $f_{|A}$ is a topologically transitive homeomorphism.* □

Corollary 2.6.6 *Let A be a stable ω-limit set with infinitely many components. Then for all $x \in A$, $\omega(x, f)$ intersects all components of A.* □

The statement corresponding to the generalization of Corollary 2.3.8, dealing with continuous maps f of a compact interval, may be strengthened in the following way.

Corollary 2.6.7 *Let A be a stable ω-limit set for a continuous map $f : I \to I$ of the compact interval I. Then:*

1. *A is either a periodic orbit, a finite cycle of intervals, or a Cantor set on which f is conjugate to an adding machine;*
2. *$f_{|A}$ is topologically transitive and $A \subset \overline{\mathrm{Per}(f)}$;*
3. *f displays sensitivity to initial conditions in A if and only if A is a cycle of intervals.*

We say that f displays *sensitivity to initial conditions in an invariant set* A if there is a set $Y \supset A$ of positive (Lebesgue) measure and an $\epsilon > 0$ such that for every $x \in Y$ and every $\delta > 0$ there is a point y which is δ-close to x and an integer $m > 0$ such that $|f^m(x) - f^m(y)| > \epsilon$. See Guckenheimer [39], Dellnitz *et al.* [25].

Proof. (1) is immediate from Theorem 2.6.1. The statements in (2) are trivial for a periodic orbit and are a consequence of Theorems 2.6.1 and 2.4.1 respectively if A has infinitely many components. If A is a cycle of intervals, let $A = \omega(x)$ for $x \in I$. Given $x_0 \in \text{int}A$, there is a subsequence $\{n_j\}$ such that $f^{n_j}(x) \to x_0$. In particular there is $N > 0$ with $f^N(x) \in \text{int}A$. By invariance of ω-limit sets (Proposition 1.2.4), it follows by setting $y = f^N(x)$ that $\omega(y) = A$. But $y \in A$, which establishes transitivity. The property $A \subset \overline{\text{Per}(f)}$ follows in this case from Theorem 3.1 in Dellnitz *et al* [25]. This result also implies that a transitive cycle of intervals displays sensitivity to initial conditions, so to show (3) it is enough to show that the two first cases do not display sensitivity. By Lemma 1.3.8 we may, for any ϵ, construct an invariant compact neighbourhood of A whose components have diameter smaller than ϵ and are cycled (trivially in the periodic orbit case, according to Theorem 2.6.1 in the adding machine case). The iterates $f^n(x)$ and $f^n(y)$ of any two points x and y in the same connected component of this neighbourhood remain inside a single component for every n, hence their distance is always smaller than ϵ. This shows there is no sensitivity in A. □

2.7 Classification of Adding Machines

It is clear from the proof of Theorem 2.3.1 that the base $\underline{k}$ of the map $\alpha_{\underline{k}}$ to which topological conjugacy of $\tilde{f}$ is assured is not uniquely determined by $\tilde{f}$. A different choice of the U, W in the proof may lead in general to a different base $\underline{k}'$ and a different adding machine $\alpha_{\underline{k}'}$. However, Theorem 2.3.1 implies that $\alpha_{\underline{k}}$ and $\alpha_{\underline{k}'}$ are topologically conjugate, so in fact $\tilde{f}$ determines a unique conjugacy class of adding machines. Thus the question of classification of adding machines up to topological conjugacy arises. This section solves this classification problem, showing that the conjugacy class is determined by the prime factors of the elements of the sequence $\underline{k}$ and their multiplicities. We first recall some general definitions and results for later use.

Definition 2.7.1 Let f be a homeomorphism of a compact metric space X. We say that $\varphi \in C(X, \mathbb{C})$ is an *eigenfunction* of f with *eigenvalue* λ if for all $x \in X$ we have $\varphi \circ f(x) = \lambda\varphi(x)$. □

It is well known that if f is topologically transitive then all the eigenvalues are simple and lie on the unit circle. The set of all eigenvalues of f forms a countable multiplicative subgroup of the unit circle, and if φ is an eigenfunction then $|\varphi(x)|$ is constant, see Walters [118]. Hence, without loss of generality, we may work only with eigenfunctions such that $|\varphi(x)| = 1$ for all $x \in X$.

Definition 2.7.2 Let f be a homeomorphism of a compact metric space X. We say that f has *topologically discrete spectrum* if the smallest closed linear subspace of $C(X,\mathbb{C})$ containing the eigenfunctions of f is $C(X,\mathbb{C})$. □

We then have:

Proposition 2.7.3 *Two minimal homeomorphisms of compact metric spaces, both having topologically discrete spectra, are topologically conjugate if and only if they have the same eigenvalues.*

Proof. See Walters [118]. □

Given $\alpha_{\underline{k}} : \Sigma_{\underline{k}} \to \Sigma_{\underline{k}}$, denote by $E_{\underline{k}}$ the set of eigenvalues of $\alpha_{\underline{k}}$. Since $\alpha_{\underline{k}}$ is minimal, it is transitive, so $E_{\underline{k}}$ is a countable multiplicative subgroup of the unit circle. The next lemma shows that $\alpha_{\underline{k}}$ has topologically discrete spectrum, showing $E_{\underline{k}}$ to be a complete topological invariant for $\alpha_{\underline{k}}$.

Lemma 2.7.4 *Let $\alpha_{\underline{k}} : \Sigma_{\underline{k}} \to \Sigma_{\underline{k}}$ be an adding machine. Then $\alpha_{\underline{k}}$ has topologically discrete spectrum and*

$$E_{\underline{k}} = \{\lambda \in \mathbb{C} : \lambda = e^{2\pi i m/k_1 \dots k_n},\ 0 \le m < k_1 \dots k_n \text{ for some } n > 0\}.$$

Proof. We first prove the second assertion. Note that the set of eigenfunctions is closed under multiplication.

Let $n > 0$ and set $E_n = \{\lambda \in \mathbb{C} : \lambda = e^{2\pi i m/k_1 \dots k_n},\ 0 \le m < k_1 \dots k_n\}$. Let $\underline{x} = (x_1, x_2, \dots) \in \Sigma_{\underline{k}}$. For convenience let $w = e^{2\pi i/k_1 \dots k_n}$. Define

$$\varphi_n(\underline{x}) = \begin{cases} 1 & \text{if } (x_1,\dots,x_n) = (0,0\dots,0) \\ w & \text{if } (x_1,\dots,x_n) = (1,0,\dots,0) \\ \vdots & \vdots \\ w^{k_1-1} & \text{if } (x_1,\dots,x_n) = (k_1-1,0,\dots,0) \\ w^{k_1} & \text{if } (x_1,\dots,x_n) = (0,1,0,\dots,0) \\ w^{k_1+1} & \text{if } (x_1,\dots,x_n) = (1,1,\dots,0) \\ \vdots & \vdots \\ w^{k_1 k_2 \dots k_n - 1} & \text{if } (x_1,\dots,x_n) = (k_1-1,k_2-1,\dots,k_n-1) \end{cases}$$

Then $\varphi_n(\underline{x})$ is an eigenfunction with eigenvalue $e^{2\pi i/k_1 \dots k_n}$. Notice that $\varphi_n(\underline{x})$ depends only on the first n coordinates of $\underline{x}$.

By considering the successive powers of φ_n, we conclude that $E_n \subset E_{\underline{k}}$, hence by induction

$$\bigcup_{n\in\mathbb{N}} E_n = \{\lambda \in \mathbb{C} : \lambda = e^{2\pi i m/k_1 \dots k_n},\ 0 \le m < k_1 \dots k_n \text{ for some } j > 0\} \subset E_{\underline{k}}.$$

We now prove the converse inclusion by showing that continuous eigenfunctions are constant on some cylinder. Suppose $\lambda = e^{2\pi i\omega}$, $\omega \in [0,1)$ is an eigenvalue corresponding to the eigenfunction φ. By continuity of φ, given any $\epsilon > 0$ we can choose $n > 0$ such that $\varphi(C_{n;i_1,\dots,i_n}) \subset e^{2\pi i[a,b]}$ with $b - a < \epsilon$ and the bounds are attained; that is, there are $\underline{x}_1, \underline{x}_2 \in C_{n;i_1,\dots,i_n}$ such that $\varphi(\underline{x}_1) = e^{2\pi i a}$, $\varphi(\underline{x}_2) = e^{2\pi i b}$. Choose $\epsilon < \frac{1}{4}$ and $n > 0$ satisfying this condition. Since $\alpha_{\underline{k}}^{k_1 \dots k_n}|_{C_{n;i_1,\dots,i_n}}$ is a homeomorphism, it follows from equation (2.2) that

$$\varphi(\alpha_{\underline{k}}^{k_1 \dots k_n}(C_{n;i_1,\dots,i_n})) = \varphi(C_{n;i_1,\dots,i_n}) \subset e^{2\pi i[a,b]}, \tag{2.19}$$

the bounds being attained. But by definition of φ we also have

$$\varphi(\alpha_{\underline{k}}^{k_1 \dots k_n}(C_{n;i_1,\dots,i_n})) = e^{2\pi i k_1 \dots k_n \omega} \varphi(C_{n;i_1,\dots,i_n}) \subset e^{2\pi i[a + k_1 \dots k_n \omega, b + k_1 \dots k_n \omega]}, \tag{2.20}$$

once again the bounds being attained. Equations (2.19) and (2.20) together imply that

$$k_1 \dots k_n \omega \equiv 0 \pmod 1. \tag{2.21}$$

Without loss of generality we take $0 \le \omega < 1$. If $\omega \in \mathbb{R} \setminus \mathbb{Q}$ there is no n satisfying (2.21), so $e^{2\pi i \omega}$ cannot be an eigenvalue. If $\omega \in \mathbb{Q}$, say $\omega = p/q$ with p, q coprime, (2.21) implies $p/q = m/k_1 \dots k_n$ for some $m < k_1 \dots k_n$. Thus $\lambda \in E_n$ for some n, which shows that $E_{\underline{k}} \subset \bigcup_{n \in \mathbb{N}} E_n$.

Let $\mathcal{A}$ be the subalgebra of $C(\Sigma_{\underline{k}}, \mathbb{C})$ generated by the eigenfunctions, that is, the collection of all finite linear combinations of eigenfunctions. Then $\mathcal{A}$ contains the constant functions, is closed under complex conjugation, and separates points. By the Stone-Weierstrass Theorem the closure of $\mathcal{A}$ is $C(\Sigma_{\underline{k}}, \mathbb{C})$. By Definition 2.7.3 f has topologically discrete spectrum. □

Remark 2.7.5 Another proof that $E_{\underline{k}} \subset \bigcup_{n \in \mathbb{N}} E_n$ may be given along the following lines, as pointed out to us by Peter Walters. The system $(\Sigma_{\underline{k}}, \alpha_{\underline{k}}, \mathcal{B})$, where $\mathcal{B}$ is the Borel σ- algebra of $\Sigma_{\underline{k}}$, is uniquely ergodic, the unique invariant measure μ being as constructed in §2.2. The eigenfunctions φ_n defined in the first part of the proof are $L^2(\mu)$-orthonormal; and since the characters of $(\Sigma_{\underline{k}}, \alpha_{\underline{k}})$ as a topological group are exactly the φ_n, they form an orthonormal basis for $L^2(\mu)$, hence span $C(\Sigma_{\underline{k}}, \mathbb{C})$. □

We now deduce a simpler combinatorial invariant from Lemma 2.7.4. Given a sequence of integers $\underline{k}$ with $k_n \ge 2$ for all n, call a prime p a *prime factor* of $\underline{k}$ if p is a prime factor of some k_i. Define the *multiplicity* $m(p)$ of a prime factor p as the sum of the powers to which p occurs as a factor in all the k_i, allowing $m(p)$ to be infinite if p occurs in infinitely many k_i. Let $\mathbb{P}$ denote the set of all primes. Then $m : \mathbb{P} \to \mathbb{N} \cup \{\infty\}$ is the *multiplicity function* of $\underline{k}$.

Theorem 2.7.6 *Let $\underline{k}$ and $\underline{l}$ be sequences of integers ≥ 2. Then $\alpha_{\underline{k}}$ and $\alpha_{\underline{l}}$ are topologically conjugate if and only if $\underline{k}$ and $\underline{l}$ have the same multiplicity function.*

Proof. The group E_k is periodic abelian, hence isomorphic to the direct sum of its p-components $E_k(p)$ by Fuchs [31]. Here the subgroup $E_k(p)$ is the set of elements of order p^r for some $r \in \mathbb{N}$. By Lemma 2.7.4 it is easy to see that $E_k(p) \simeq \mathbb{Z}_{p^{m(p)}}$, where $\mathbb{Z}_{p^\infty}$ denotes the *Prüfer* or *quasicyclic* group of all p^rth roots of unity for $r \in \mathbb{N}$, see Fuchs [31]. Since isomorphic abelian groups have isomorphic p-components, the result follows. □

We deduce that every adding machine is conjugate to a *prime* adding machine, that is, one in which each k_j is prime:

Corollary 2.7.7 *Let $k_i = p_{i_1} \dots p_{i_{j_i}}$ be the prime factorization of k_i, $i \geq 1$. Then $\alpha_{\underline{k}}$ is topologically conjugate to $\alpha_{\underline{p}}$, where*

$$\underline{p} = (p_{1_1}, \dots, p_{1_{j_1}}, p_{2_1}, \dots, p_{2_{j_2}}, \dots, p_{i_1}, \dots, p_{i_{j_i}}, \dots)$$

Proof. Their eigenvalue groups $E_{\underline{k}}$ and $E_{\underline{p}}$ coincide. □

Remark 2.7.8 From Theorem 2.7.6 we conclude, for instance, that the adding machines with bases $\underline{k} = (2, 3, 2, 3, \dots)$ and $\underline{l} = (3, 2, 3, 2, \dots)$ are conjugate. More surprisingly, they are both conjugate to the adding machine defined by $(2, 2, 3, 2, 2, 3, \dots)$, or indeed to any adding machine whose base contains only the primes 2 and 3, each of them with infinite multiplicity. □

Remark 2.7.9 The preceding methods also allow us to classify Denjoy maps up to topological conjugacy. Given a Denjoy map D_ω with minimal Cantor set K_ω, semiconjugacy to irrational rotation by ω ensures that D_ω has topologically discrete spectrum. This fact also shows that the eigenvalue set is the infinite cyclic group $E_\omega = \{\lambda \in \mathbb{C} : \lambda = e^{2\pi i n \omega},\ n \in \mathbb{Z}\}$. Since by definition $\omega \in \mathbb{R} \setminus \mathbb{Q}$ we immediately see that (1) D_ω is not topologically conjugate to any adding machine $\alpha_{\underline{k}}$; (2) D_{ω_1} and D_{ω_2} are topologically conjugate if and only if $\omega_1 = \omega_2$. The first statement proves Corollary 2.3.9; the second furnishes another proof that the rotation number ω is a complete invariant for Denjoy maps. □

2.8 Existence of Stable Adding Machines

Let $\underline{k} = (k_1, k_2, \dots)$ be a sequence of positive integers greater than 1. We sketch the construction of a homeomorphism φ of the plane such that φ admits a stable transitive Cantor set A on which the dynamics is topologically conjugate to the adding machine $\alpha_{\underline{k}}$. The map φ leaves a disk invariant, and can be extended to a homeomorphism of the 2-sphere with a repellor at ∞. Moreover, we can make φ a diffeomorphism on the complement of A.

The construction is inductive and suggested by our constructions characterizing stable adding machines in the preceding sections. Arrange k_1 disjoint equal discs D_i, where $i = 0, \dots, k_1 - 1$, at the vertices of a regular k_1-gon. Inside each D_i place a smaller disk E_i. Define a C^∞ flow ψ_1^t on $\mathbb{R}^2 \setminus (\bigcup_{i=0}^{k_1-1} D_i)$ as in Figure 2.2.

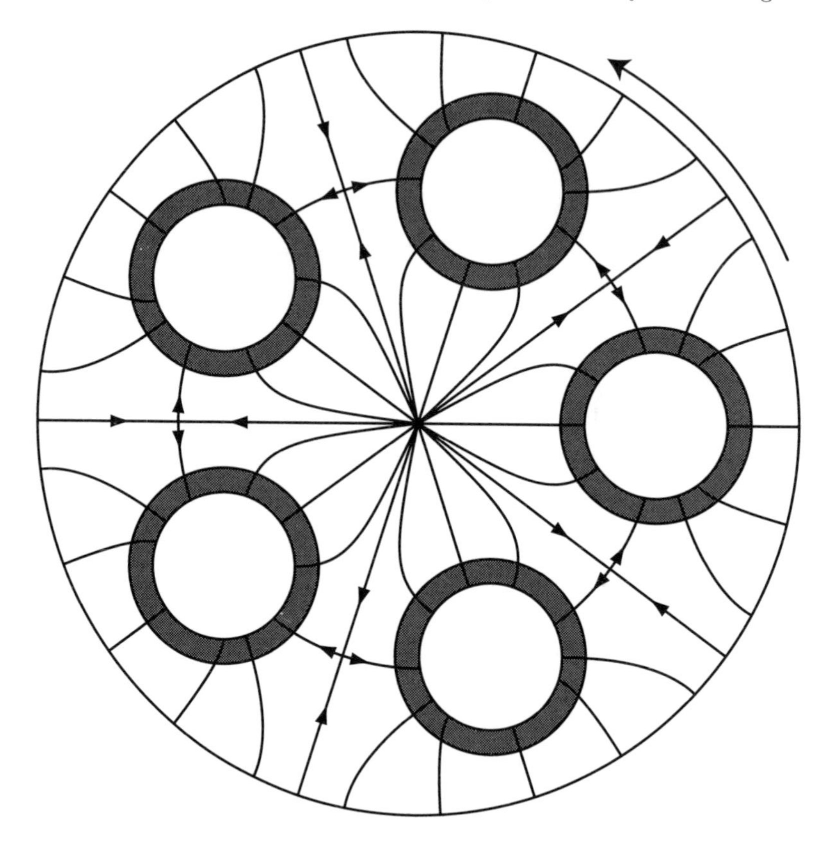

Figure 2.2: The flow ψ_1^t. The larger disks are the D_i, the smaller ones the E_i.

Parametrize t in such a way that the time-1 map ψ_1^1 maps ∂D_i to ∂E_i. Let φ_1 be ψ_1^1 composed with rotation by $2\pi/k_1$. Then φ_1 cycles the D_i; more precisely, $\varphi_1(\partial D_i) = \partial E_{i+1 \pmod{k_1}}$.

Inside each E_i place k_2 disks $D_{i_1,i_2}, i_2 = 0, \ldots, k_2 - 1$ in a k_2-gon, with k_2 smaller disks E_{i_1,i_2} inside. See Figure 2.3.

Using a similar construction to the above, extend φ_1 smoothly to a map $\varphi_2 : \mathbb{R}^2 \setminus (\bigcup_{i_1,i_2} D_{i_1,i_2}) \to \mathbb{R}^2 \setminus (\bigcup_{i_1,i_2} E_{i_1,i_2})$ such that φ_2 cycles the $k_1 k_2$ discs like a truncated adding machine on the subscripts i_1, i_2. By this we mean that

$$\varphi(D_{i_1,i_2}) = \begin{cases} E_{i_1+1,i_2} & \text{if} \quad i_1 < k_1 - 1 \\ E_{0,i_2+1 \pmod{k_2}} & \text{if} \quad i_1 = k_1 - 1. \end{cases}$$

This involves inserting at the boundary of $D_{i_1,0}$ an annular region in which a rotation of $2\pi/k_2$ is introduced at the inner edge, varying smoothly from zero rotation on the outer edge.

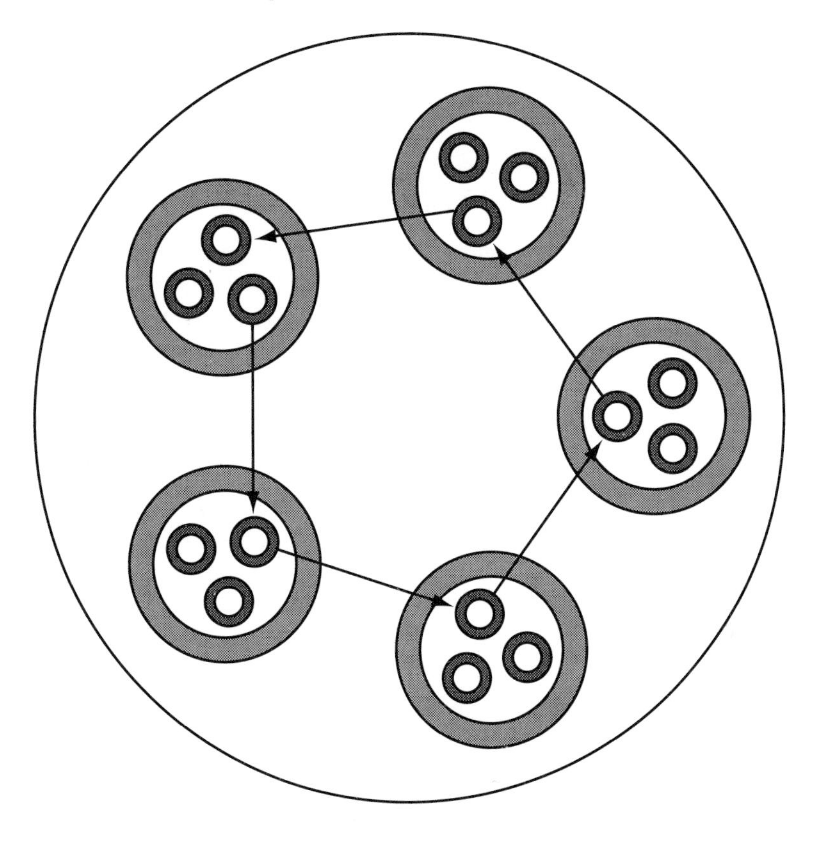

Figure 2.3: Iteration of the construction to produce adding machine dynamics.

Proceed inductively, introducing at each stage sets of k_j smaller and smaller discs for $j = 3, 4, \ldots$. Let A be the intersection of all the discs. This process defines a homeomorphism $\varphi_\infty : \mathbb{R}^2 \setminus A \to \mathbb{R}^2 \setminus A$. Extend continuously to a map $\varphi : \mathbb{R}^2 \to \mathbb{R}^2$ by making φ act on A as the adding machine $\alpha_{\underline{k}}$. By Remark 2.3.4 this construction ensures continuity of φ; indeed, φ is a homeomorphism, and away from A it is a C^∞ diffeomorphism.

Note that A is transitive since φ is minimal on A. It is also Liapunov stable, because the discs $D_{i_1,i_2,\ldots,i_n}$ are invariant under forward iterates of φ and form a basis of neighbourhoods for A.

It is easy to see directly that A is not an asymptotically stable attractor, as is implied in the abstract by Theorem 1.4.6. As implied by Remark 1.3.12, A is accumulated by ω-limit sets other than itself – in this case the infinite family of periodic saddles arising from the construction. *Every* point x except on the stable manifolds of these countably many saddles has ω-limit set $\omega(x) = A$. So for every open neighbourhood U of A we have $\omega(x) = A$ for a set of x of full Lebesgue measure; therefore A is a Milnor attractor. This also implies that φ has an SBR measure, which is exactly the unique invariant measure μ constructed from the symbolic dynamics of the adding machine $\alpha_{\underline{k}}$ in §2.2.

2.9 Historical remarks and further comments

Remark 2.9.1 (Connected components of stable sets) As mentioned in the comments to the previous chapter, Dellnitz *et al.* in their paper *The Structure of Symmetric Attractors* [25] were the first to consider implications of stability in terms of the connected components of an invariant set. Their investigation and results are based on an *ad hoc* topological lemma about the connected components of the preimage set of a stable set A. Although in a sense their results are less sharp than ours, since for instance they do not uncover the adding machine structure of stable ω-limit sets, they extend in other directions – for instance, connections with sensitivity to initial conditions.

This paper was inspirational for our work, either for the results in this chapter or for Theorem 1.4.6. This last result is to the best of our knowledge original. It was proved simultaneously to the results in this Chapter; only for convenience of exposition does it belong to a different chapter of this book.

Remark 2.9.2 (Adding machines in dynamics) The occurrence of adding machines in dynamics is not novel. In a first phase, during the 50's and 60's, they occurred mainly in classification problems in abstract topological dynamics, as examples of minimal homeomorphisms of the Cantor set; see, for instance, Gottschalk [37], Brown [19]. The approach was that of symbolic dynamics. At this stage the terminology "adding machine" was not yet standard; Gottschalk, for instance, refers to the base $\underline{2}$ adding machine as the "dyadic group" (by our results in § 2.5, adding machines are indeed conjugate to translations in a topological group). Later authors refer to it as the "odometer" [55] or the "infinite register shift" [119], but by this stage the (equivalent) terminology "adding machine" was already standard.

Interest in one-dimensional dynamics from the early 70's on was inevitably bound to expose the ubiquity of adding machines as subsets of the non-wandering set. The reason is clear from Corollary 2.6.7: stable ω-limit sets are either periodic orbits, cycles of intervals or adding machines. Thus, for instance, the adding machine structure is unavoidably present in the Feigenbaum Cantor set, and its study revealed adding machines not just as abstract dynamical systems *per se* but appearing in an essential way in "real-life" dynamical systems as interval maps. In this respect the most significant results and observations are due to Misiurewicz [72]. He considers the set of all unimodal, negative Schwarzian derivative interval maps which, for every $n > 0$, have exactly one periodic orbit of period 2^n and no other periodic points – that is, at the Feigenbaum limit of period-doubling. He then concludes that all these maps are topologically conjugate, and that for a given map f in this class the periodic points accumulate in an invariant Cantor set

C on which f is conjugate to the base $\underline{2}$ adding machine, and the non-wandering set of f is precisely $C \bigcup \overline{\text{Per}(f)}$.

Katznelson [55] studies adding machines in circle maps. He investigates the action of diffeomorphisms of the circle of irrational rotation number on Lebesgue measure. His main result is that such a map induces a specific type of adding machine in the collection of subsets of S^1 with positive Lebesgue measure.

In 1981 Jonker and Rand [52] perform a complete topological characterization of the non-wandering set for continuous unimodal maps of the interval. They prove that, given a map f in this class, there exists a canonical decomposition of the non-wandering set $\Omega(f)$ into a finite or denumerably infinite collection of compact invariant sets $\{\Omega_i\}$. In the latter case this collection includes a compact invariant set Ω_∞ which is totally disconnected and on which f is topologically semiconjugate to an adding machine. This "spectral decomposition" is constructed using the kneading theory of Milnor and Thurston [71] and is specific to unimodal maps.

Block and Coven [15] study the following problem. Given a continuous map $f : I \to I$ of a compact interval I, the set $\Lambda(f)$ of ω-limit points of f is closed. If f is piecewise monotone, then $\Lambda(f) = \overline{\text{Per}(f)}$. In the opposite case, this is not necessarily true. Block and Coven prove that, if $x \in \overline{\text{Per}(f)} \setminus \Lambda(f)$, then $\omega(x)$ is an infinite minimal set, and construct an example of such an f where $\Lambda(f) \setminus \overline{\text{Per}(f)}$ is a Cantor set where the dynamics is (conjugate to) a base $\underline{2}$ adding machine. Of course, by our results in this chapter this adding machine must be Liapunov unstable; this fact is clear from their geometric construction.

Willms [119] studies the asymptotic behaviour of iterations of piecewise monotonic maps, allowing for a finite number of jump discontinuities. He arrives at a spectral decomposition for the non-wandering set in the spirit of Jonker and Rand [52]. The main result states that the non-wandering set may include a finite number $R_1, \ldots, R_s$ of Cantor sets on each of which the dynamics is conjugate to an adding machine map. Hofbauer and Raith [49] perform a very similar construction in the same conditions.

Another line of research which independently led to the uncovering of the presence of adding machines in dynamics was sparked by a question of Smale [111] in his 1962 address to the International Congress of Mathematicians. In a review address on *Dynamical systems and the topological conjugacy problem for diffeomorphisms*, Smale asks whether it is true that a diffeomorphism of S^2 whose periodic points are all hyperbolic and the corresponding stable and unstable manifolds intersect transversely must possess at least one "elliptic point". In modern language, the problem can be stated in the following terms: does a Kupka-Smale diffeomorphism of S^2 necessarily have at least one source or one sink?

In 1976 Bowen and Franks [17] provided a negative answer to this question by constructing an example of a C^1 Kupka-Smale diffeomorphism of S^2 with neither sinks nor sources. In 1980 Franks and Young [30] improved the differentiability class to C^2. In both cases the geometric construction is essentially the same, and quite closely related to our own construction in § 2.8. The main step is to produce a Kupka-Smale embedding from a disk D^2 in the plane to itself; this is achieved by iterated surgeries which inductively transform the period 2^n sinks into a pair of period 2^n saddles and period 2^{n+1} sinks. In the limit, a Kupka-Smale is obtained which contains saddles of all periods of the form 2^n and a base $\underline{2}$ adding machine, in precisely the same way as in § 2.8.

In 1989 Gambaudo, van Strien and Tresser [33] apply renormalization group methods instead of surgery to improve the differentiability of these examples to C^∞. In this paper the authors observe, as we do in § 2.8, that the unique invariant measure on the base $\underline{2}$ adding machine thus constructed is an SBR measure. However, their approach is not constructive, and geometric intuition is lost. In 1992 Gambaudo and Tresser [35] return to the constructive approach and again using the main geometric idea on the d-dimensional disk D^d, construct $C^{r(d)}$ diffeomorphisms, showing that $r(d)$ is a strictly increasing function of d which tends to infinity with d.

In 1993 Barge and Walker [7], motivated by this line of research, study a problem which is also closely related to ours: the embedding of non-wandering fibres as structures over an adding machine. They show that no homeomorphism of the plane may possess non-wandering Knaster continua on which the homeomorphism is semi-conjugate to an adding machine, and provide a counterexample showing this to be possible in three or more dimensions.

Remark 2.9.3 (Conjugacy of systems with stable adding machines) It must be stressed that Theorem 2.7.6 gives a classification of *adding machines* up to conjugacy. In the original dynamical system (X, f) this means that we may construct a homeomorphism of the adding machine Cantor set A which conjugates $f_{|A}$ with any other adding machine in the same conjugacy class, as given by Theorem 2.7.6. However, this homeomorphism will *not* in general extend to a homeomorphism of a neighbourhood of the Cantor set A. The problem is similar in spirit to that of extending a homeomorphism of a knot into another to homeomorphisms of tubular neighbourhoods of these knots. Thus, for instance, it will in general impossible to conjugate maps of the interval admitting stable adding machines of base $\underline{k_1} = (2, 3, 2, 2, 2, \dots)$ and $\underline{k_2} = (3, 2, 2, 2, 2, \dots)$ – even though the adding machines themselves are conjugate – since the former will necessarily have periodic points of periods $2, 6, 12, \dots$ and the latter of periods $3, 6, 12, \dots$.

It is possible and interesting to consider a classification theorem for the conjugacy classes of maps admitting stable adding machines. The first step would involve finding the suitable topological conditions which the ambient space would have to satisfy to make that classification possible. This problem lies outside the scope of the present work.

Another interesting problem along these lines is whether the above analogy between the extension of homeomorphisms to the ambient space in this context and in the context of knot theory is just an analogy or has a deeper meaning. It is likely that by suspension of an adding machine diffeomorphism one might translate this problem into the language of knots and links and so be able to produce rigourous statements. Recent work follows this line of research. Alsedà *et al.* [2] study the sequence of average linking numbers induced by the braiding associated to such a construction and show that this sequence must converge. Gambaudo, Sullivan and Tresser [34] associate to an orientation-preserving diffeomorphism of a disk an interval of topological invariants, each point of which describes the way an infinite sequence of periodic orbits are asymptotically linked one around the other; they explicitly describe and analyse the adding machine case.

Remark 2.9.4 (Persistence of adding machines) Stable adding machines can *never* be Axiom A, or even hyperbolic: Axiom A attractors are asymptotically stable and thus by Theorem 1.4.6 have finitely many components, whereas saddle-like hyperbolic sets (that is, such that $A \neq \overline{\bigcup_{x \in A} W^u(x)}$) must be unstable. Hence stable adding machines cannot be uniformly hyperbolic. In view of the Stability Conjecture [64], this implies that stable adding machines are structurally unstable, at least in the case where the original map f is a diffeomorphism. However, there is still the question of persistence: experience with one-dimensional unimodal maps suggests that the Feigenbaum adding machine, although clearly structurally unstable, should be persistent (in the sense of existing in a residual set of an open subset) in an adequate class of maps.

Remark 2.9.5 (Existence of stable adding machines) At the time of first writing these results, it seemed important to establish that adding machines of arbitrary base sequence do indeed exist in dynamics, so that our results are non-vacuous. This is the purpose of our example in § 2.8. After developing this particular piece of work it was pointed out by J. Guckenheimer [38] that all stable adding machines are present in the bifurcation diagram of generic 1-parameter families of unimodal maps of the interval. The construction is, in a sense, sketched in Milnor [69].

The geometrical construction in the plane in § 2.8 suggests itself from the proof of Theorem 2.6.1, and was thought of as an *ad hoc* example. However, it has been pointed out above that this construction is closely related to others in the literature, namely those of Bowen and Franks [17], Franks and Young [30], Gambaudo and Tresser [35]. At the point of conception of this work, the questions of how common stable adding machines are in dynamics, the topological conditions for their existence and possible topological obstructions on admissible base sequences seemed open. The example in § 2.8 was constructed without prior knowledge of these sources. Moreover, it illustrates the non-trivial fact that *every* base sequence is admissible for a stable adding machine.

Remark 2.9.6 (Adding machines and periodic points) Theorem 2.4.1 clearly asks for immediate generalization. In 1993 J. M. Gambaudo asked whether it was true that stable adding machines were always contained in $\overline{\mathrm{Per}(f)}$. This question provided the motivation for Theorem 2.4.1 and for unsuccessful attempts at a general result. We observed in Remark 2.4.4 that a necessary condition for such a statement is that A be itself a Cantor set. However, a general result of this nature proved elusive.

In 1995 Bell and Meyer [10] proved the analog of Theorem 2.4.1 for homeomorphisms of the plane, again using a fixed point theorem which works specifically in dimension 2 – the Bell-Cartwright-Littlewood theorem, see [8], [9]. They also constructed an example of a homeomorphism of $\mathbb{R}^3$ which, restricted to a solid torus, possesses a stable adding machine and *no periodic points*. The ω-limit sets which by virtue of Remark 1.3.12 must accumulate on the adding machine are in this case a countable collection of unstable invariant tori. These results settle Gambaudo's original question in a somewhat surprising way.

Remark 2.9.7 (Organization of results) We would like to finish this chapter with a note on its organization. The reader may have noticed that there is a certain extent of overlap and sometimes redundancy between some of its sections. To be specific: in § 2.3 we prove Theorem 2.3.1 for stable transitive sets, while in § 2.5 we prove it for stable ω-limit sets – a strictly stronger result. In § 2.3 the construction of invariant partitions is made *ab initio* from the definition of Liapunov stability; a more economical way is to use Lemma 1.3.8 which yields such partitions easily, and which is exactly what is done in § 2.5 – and makes the result work for stable ω-limit sets, not necessarily transitive. Furthermore, all the results of § 2.4 are most easily proved by applying the methods of § 2.6 to the special case of interval maps: compact connected invariant sets in $\mathbb{R}$ are just compact intervals, which have the fixed point property. It is meaningful to ask why we did not choose a more economical and efficient organization of our results.

As described in the Preface, we chose to organize this book in a way that mirrors the surfacing of results. Below we summarize their evolution.

By mid-1993 the results in Lemma 1.3.13, Theorem 1.3.14, Remark 1.3.12, Theorem 1.4.6, § 2.3, § 2.7 and § 2.8 were known. They formed a coherent whole and were the core of a paper submitted to *Ergodic Theory and Dynamical Systems*. A. Manning then remarked on the similarity between our example and that of Bowen and Franks [17], thus first drawing our attention to the second line of research described in Remark 2.9.2. The elementary but crucial property of Liapunov stability in Lemma 1.3.8 was then proved. Discussions with I. Melbourne triggered attempts to generalize our results to stable ω-limit sets. By now Lemma 1.3.8 and the results leading to Remark 1.3.7 provided all the ingredients for an inverse limit approach.

Chapter 3
From Attractor to Chaotic Saddle: a journey through transverse instability

3.1 Introduction

Although not a generic situation in the broadest sense, many smooth dynamical systems defined by flows or maps leave certain submanifolds of the phase space invariant. This situation can become generic if appropriate constraints are imposed – for instance, if the system has a symmetry. However, there are many other situations in which this is the case, such as in evolutionary dynamics, synchronization of chaotic oscillators and systems with hidden symmetries.

To be specific, suppose the phase space M is a smooth finite-dimensional manifold and $f : M \to M$ is a smooth map leaving a lower dimensional submanifold N invariant – that is, $f(N) \subset N$. The restriction $g = f_{|N} : N \to N$ determines a discrete dynamical system in its own right. We address the following problem. Suppose $A \subset N$ is an attractor for g. Under what conditions is A an attractor for f on M?

Liapunov exponents [80] provide useful quantitative indicators of asymptotic expansion and contraction rates in a dynamical system. They therefore have a fundamental bearing upon questions of stability and bifurcation. In particular, local stability of fixed points or periodic orbits is determined by whether the real part of the eigenvalues are inside the unit circle or not, or equivalently by whether the corresponding Liapunov exponents are negative or not. Suppose that a hyperbolic periodic orbit P lies on an invariant submanifold N as above, and that P is attracting for $g = f_{|N}$ – that is, all its Liapunov exponents are negative regarding it as an object of N. In this case, asymptotic stability of P in the global phase space M is determined by the remaining Liapunov exponents. Specifically, if all the remaining exponents are negative, P is asymptotically stable, whereas if there is at least one positive exponent, it is a saddle.

For invariant sets with more complex dynamics, the situation is much more subtle, but the considerations above provide useful guidelines for their analysis.

One of the main goals of the present work is to show that local dynamic stability of chaotic attractors in invariant submanifolds may be described in terms of their *normal* Liapunov exponents – that is, the extra exponents that are introduced when considering the attractor as a subset of the global phase space M instead of just the invariant manifold N. A semi-rigorous approach to this question is given by the following argument. Suppose dim $M = m$, dim $N = n < m$. Given any ergodic invariant measure μ supported in A, then for μ-almost all $x \in A$ there exist $m - n$ *normal Liapunov exponents* $\lambda_\perp^i(x)$ (that is, Liapunov exponents whose corresponding eigenspaces are not tangent to N). By ergodicity they are constant μ-a.e., allowing us to drop the dependence on x. If all $\lambda_\perp^i < 0$, then by standard results [102] we know that there exists a set $B_\mu \subset A$ of full μ-measure such that for all $x \in B_\mu$ there exists an $m - n$-dimensional local stable manifold $W_{loc}^s(x)$ which is transverse to N. Therefore, if we consider a neighbourhood U of A in M, points lying in the intersection $\bigcup_{x \in B_\mu} W_{loc}^s(x) \cap U$ will be forward-asymptotic to A. On the other hand, if at least one of these normal Liapunov exponents is positive, then for all $x \in B_\mu$ there exists a corresponding unstable manifold and A must be Liapunov unstable.

However, this is far from being the end of the story. Indeed, invariant measures on A are seldom unique; for instance, associated to each periodic orbit contained in an attractor is a Dirac ergodic measure whose support is the orbit. Ergodic measures are thus not unique if A contains more than one periodic orbit – as is the case with chaotic attractors, including all (non-trivial) Axiom A attractors. Each ergodic measure carries its own Liapunov exponents, so the question of stability in transverse directions arises independently for *every* ergodic measure supported in A. For example, two different periodic orbits in A may have normal exponents of different signs. If there exists a 'natural' measure on A, then Lebesgue-almost all points have corresponding normal exponents and manifolds, but there can still be a dense set in A with the opposite behaviour.

Liapunov exponents only give a linearized picture of stability. The 'global' stability of an attractor will be typically determined by dynamics far from A, and cannot be found by a local analysis of higher order terms in some Taylor expansion. As noticed by Ott and Sommerer [83], there are essentially two different types of global dynamics: intermittency, where the local unstable manifolds fold back on A (giving rise to the occurrence of 'fluctuations' away from A that are forced back again by global dynamics) or riddled basins (see Alexander *et al.* [1]), where the dense unstable manifolds are contained in the basin of a second, distinct attractor. We generalize these concepts by defining a 'locally riddled basin', which includes the case where the basin of A is open but local normal unstable manifolds exist in a dense set in A.

This chapter is organized as follows. In § 3.1.1 we introduce some definitions and terminology. § 3.2 describes the local theory for a manifold M with an embedded invariant submanifold N on which $f_{|N}$ has an asymptotically stable attractor A. We give a characterization of the local normal stability of A in terms of the spectrum of normal Liapunov exponents. In § 3.3 we develop an appropriate bifurcation theory and show that the relevant bifurcations arise in a persistent way under additional assumptions on the dynamics on A. § 3.4 considers two numerical examples showing the same local behaviour but contrasting global behaviour; these examples illustrate much of the theory in § 3.2 and § 3.3. An application to synchronization of two coupled identical chaotic systems, illustrated by an electronic experiment, is presented in § 3.5. Finally, in § 3.6 we outline the short history of the problem, discuss the notion of criticality for the bifurcations studied, consider the effect of low levels of noise on the attractors and indicate some further directions of study. For ease of reference the figures are grouped at the end of the chapter.

3.1.1 Riddled and locally riddled basins

Let M be an m-dimensional Riemannian manifold and let $f : M \to M$ be a smooth map. We denote Lebesgue measure on M – which may, for instance, be derived from a volume form – by $\ell(\cdot)$. We say a compact transitive set A is *chaotic* if it has sensitive dependence on initial conditions, see Guckenheimer [39].

The following concept was introduced by Alexander *et al.* [1] to deal with the case where the basin of a Milnor attractor contains no open sets.

Definition 3.1.1 *A Milnor attractor A has a* riddled basin *if for all $x \in \mathcal{B}(A)$ and $\delta > 0$ we have*

$$\ell\left(B_\delta(x) \cap \mathcal{B}(A)\right) \ell\left(B_\delta(x) \cap \mathcal{B}(A)^c\right) > 0. \tag{3.1}$$

If there is another Milnor attractor C such that $\mathcal{B}(A)^c$ in equation (3.1) may be replaced with $\mathcal{B}(C)$, then we say that the basin of A is riddled with the basin of C. If $\mathcal{B}(A)$ and $\mathcal{B}(C)$ are riddled with each other, we say they are intertwined.

We next offer a variation of this definition which allows for the possibility of A having an open basin but *not* being Liapunov stable. Given a Milnor attractor A and an open neighbourhood V of A, set $U(V) = \bigcap_{n\geq 0} f^{-n}(V)$; that is, $U(V)$ is the set of points in V whose iterates always remain in V.

Definition 3.1.2 *A Milnor attractor A has a* locally riddled basin *if there exists a neighbourhood V of A such that, for all $x \in A$ and $\delta > 0$,*

$$\ell\left(B_\delta(x) \cap U(V)^c\right) > 0.$$

This definition states that an arbitrarily small ball centered around *any* point of the attractor contains a set of positive measure which eventually leaves V under iteration. It is a broader definition than that of a riddled basin since it allows the set which is locally repelled from A to eventually 'fold back' upon A. Thus it may happen in this case that $\mathcal{B}(A)$ contains an open neighbourhood of A. Note, however, that an attractor with a locally riddled basin is *never* Liapunov stable. Thus even in the case where $\mathcal{B}(A)$ is open, A is not asymptotically stable.

A chaotic invariant transitive set A is called a *chaotic saddle* (Nusse and Yorke [79]) if there is a neighbourhood U of A such that $\mathcal{B}(A) \cap U \neq A$ but $\ell(\mathcal{B}(A)) = 0$.

In the next definition we suppose that $N \supset A$ is a forward-invariant n-dimensional submanifold of M, where $n < m$.

Definition 3.1.3 *We say that A is a* normally repelling chaotic saddle *if $\mathcal{B}(A) \neq A$ and $\mathcal{B}(A) \subset N$.*

A normally repelling chaotic saddle is thus an extreme example of a chaotic saddle.

All the measures we work with, as stated in Chapter 1, are Borel measures – that is, measures defined on the Borel σ-algebra. Given a non-empty compact set A invariant under a continuous map f, we denote by $\mathcal{M}_f(A)$ and $Erg_f(A)$ respectively the sets of invariant probability measures and ergodic measures supported in A; by Remark 1.2.15 both $\mathcal{M}_f(A)$ and $Erg_f(A)$ are non-empty.

3.1.2 Symmetry and invariant submanifolds

Symmetry provides a setting where invariant subspaces and invariant submanifolds arise naturally. Suppose that f commutes with the smooth action of a (compact Lie) group of symmetries Γ on M, and let $\Sigma \leq \Gamma$ be a subgroup. Then the fixed-point submanifold

$$N = \mathrm{Fix}(\Sigma) = \{x \in M \ : \ \sigma(x) = x \ \text{ for all } \sigma \in \Sigma\}$$

is invariant under f. The states $x(t)$ that lie in N are those for which, at every instant of time, $x(t)$ is invariant under all elements of Σ. If the dynamics of $A \subset N$ is chaotic, then states of the system that lie in A are 'spatially' ordered (have symmetry group Σ) but are temporally chaotic. If A loses transverse stability, breaking the spatial symmetry but leaving the dynamics chaotic, we have a transition to 'spatio-temporal chaos'. An example for Bénard convection in an annular geometry may be found in Caponeri and Ciliberto [21]. This is perhaps the simplest manner in which such a transition can occur.

In § 3.4 and § 3.5 we consider examples of invariant subspaces caused by dynamics with $\mathbb{Z}_2$ symmetry. In § 3.5 we study an experimental system of two coupled identical oscillators. In this case the symmetry group $\mathbb{Z}_2$ generated by the permutation of the oscillators means that the invariant fixed-point space is the space of synchronized states – that is, that on which the instantaneous states of both oscillators coincide.

For many groups, the group action will typically stratify the manifold into a hierarchy of fixed point subspaces corresponding to the isotropy types of points on the manifold. This will give rise to a hierarchy of invariant manifolds of differing dimensions, and there is the possibility that loss of transverse stability of an attractor in a low dimensional manifold will proceed by losing stability into progressively higher dimensional invariant subspaces in succession. In this work we will only consider the case of one invariant subspace.

Although symmetry is the most obvious way for dynamical systems to leave specific submanifolds invariant, there are other natural ways for this to happen. Rand *et al.* [96] study ecological models in which the subspace N represents zero population for some particular phenotype. Because a species with zero population cannot reproduce, this space must be invariant. Loss of transverse stability of A is now an 'invasion' of the ecology by a new species, triggered by a small perturbation to non-zero population values – possibly through a mutation. Another example is given by Kocarev *et al.* [57] who consider linear forcing of one system by another identical system. Although the symmetry is destroyed by the unidirectional coupling, there is still an invariant subspace of synchronous states.

3.2 Normal Liapunov exponents and stability indices

We define the Liapunov exponents (also known as characteristic exponents, see Ruelle [104]) and discuss the role of the normal Liapunov exponents in determining the stability of the attractor A for $f_{|N}$ with respect to perturbations outside N. These exponents are a.e. constant for each ergodic measure supported on A, and we refer to the set of all normal Liapunov exponents as the *normal spectrum* of A. Lemma 3.2.8 shows, under an appropriate assumption, that the upper and lower limits of the normal spectrum can be characterized by real numbers $\lambda_{\min}$ and $\Lambda_{\max}$. Theorem 3.2.12 shows that A is asymptotically stable for $\Lambda_{\max} < 0$ and Liapunov unstable for $\Lambda_{\max} > 0$. The normal spectrum also determines whether a chaotic saddle is normally repelling or not: see Theorem 3.2.16. If in addition there exists a strong SBR measure, its normal Liapunov exponents determine whether A is a Milnor attractor or a chaotic saddle: see Theorems 3.2.17 and 3.2.18.

The results of this section are collected together in Proposition 3.2.20 which classifies the stability or instability of A according to its normal spectrum.

3.2.1 Normal Liapunov exponents

Let M be an m-dimensional Riemannian manifold and let $f : M \to M$ be a $C^{1+\alpha}$ map. Suppose that f leaves an n-dimensional embedded submanifold $N \subset M$ invariant, where $n < m$ – that is, $f(N) \subseteq N$. It follows that for $p \in N$

$$d_p f(T_p N) \subseteq T_{f(p)} N. \tag{3.2}$$

Define a smooth splitting of the tangent bundle $TM = \bigcup_{p \in M} T_p M$ in a neighbourhood of N which coincides with $T_p M = T_p N \oplus (T_p N)^\perp$ when $p \in N$, the orthogonal complement $(T_p N)^\perp$ being taken with respect to the Riemannian structure in M. This is possible because N is an embedded submanifold.

We suppose throughout this chapter that $A \subset N$ is an asymptotically stable attractor for $f_{|N}$.

For $p \in A,\ 0 \neq v \in T_pN$, the *Liapunov exponent* $\lambda(p, v)$ *at the point* p *in the direction of* v is defined to be

$$\lambda(p, v) = \lim_{n\to\infty} \frac{1}{n} \log \| d_p f^n(v) \|_{T_{f^n(p)}N} \tag{3.3}$$

if this limit exists.

Given an f-invariant measure $\mu \in \mathcal{M}_f(A)$, the multiplicative ergodic theorem of Oseledec (see, for instance, Walters [118]) implies that the limit in (3.3) exists for $\mu - a.a.\ \ p \in N$ and every $v \in T_pN$.

In what follows, p will denote an arbitrary point of N. We write

$$TM_n = T_{f^n(p)}M \tag{3.4}$$

to alleviate the notation.

Given the splitting $TM_n = TN_n \oplus (TN_n)^\perp$, the derivative $d_pf : TM_0 \to TM_1$ block-decomposes in matrix form with respect to these subspaces as

$$\begin{pmatrix} d_pf \circ \Pi_{TN_0} & \Pi_{TN_1} \circ d_pf \circ \Pi_{(TN_0)^\perp} \\ 0 & \Pi_{(TN_1)^\perp} \circ d_pf \circ \Pi_{(TN_0)^\perp} \end{pmatrix} \tag{3.5}$$

where Π_V denotes the orthogonal projection onto the vector subspace V and use is made of (3.2).

Lemma 3.2.1 *With the above notation,*

$$\Pi_{(TN_n)^\perp} \circ d_pf^n \circ \Pi_{(TN_0)^\perp} =$$
$$(\Pi_{(TN_n)^\perp} d_{f^{n-1}(p)} f) \circ (\Pi_{(TN_{n-1})^\perp} d_{f^{n-2}(p)} f) \circ \ldots \circ (\Pi_{(TN_1)^\perp} d_pf \circ \Pi_{(TN_0)^\perp}).$$

Proof. This results by a straightforward computation using (3.5), the chain rule and the fact that $\Pi^2 = \Pi$ since Π is an orthogonal projection. □

Definition 3.2.2 *Consider* $p \in A,\ v \in T_pM$. *Define the* tangent Liapunov exponent at p in the direction of v *to be*

$$\lambda_{\|}(p, v) = \lim_{n\to\infty} \frac{1}{n} \log \|\Pi_{(TN_n)} \circ d_pf^n \circ \Pi_{TN_0}(v)\|_{TM_n}. \tag{3.6}$$

Similarly define the normal Liapunov exponent at p in the direction of v *to be*

$$\lambda_{\perp}(p, v) = \lim_{n\to\infty} \frac{1}{n} \log \|\Pi_{(TN_n)^\perp} d_pf^n \circ \Pi_{(TN_0)^\perp}(v)\|_{TM_n}. \tag{3.7}$$

Both definitions apply only when the limit exists.

Notice that the projection Π_{TN_n} in (3.6) is in fact the identity operator in view of (3.2), and could therefore be omitted. It is included to stress the similarity between both definitions.

Theorem 3.2.3 *Let μ be an f-invariant measure supported in A. Then, for μ-a.a. $p \in A$ and every $v \in T_pM$ the following hold:*

1. $\lambda_{\|}(p,v)$ *and* $\lambda_{\perp}(p,v)$ *exist.*
2. *If* $w = \Pi_{T_pN}(v) \neq 0$, *then* $\lambda_{\|}(p,v)$ *equals the Liapunov exponent* $\lambda(p,w)$ *for* A *considered as an attractor for* $f_{|N} : N \to N$.
3. *If* $0 \neq \Pi_{(T_pN)^{\perp}} v$, *then there exists* $s(p) \leq m-n$ *such that* $\lambda_{\perp}(p,v)$ *takes one of the values*
$$\lambda_{\perp}^1(p) < \lambda_{\perp}^2(p) < \ldots < \lambda_{\perp}^{s(p)}(p).$$
4. *The function* $p \mapsto s(p)$ *is* f*-invariant and* μ*-measurable.*
5. *There exists a filtration*
$$\{0\} = V^0(p) \subset V^1(p) \subset \ldots \subset V^{s(p)}(p) = (T_pN)^{\perp}$$
with
$$\Pi_{(T_pN)^{\perp}} \circ d_pf(V^i(p)) \subset V^i(f(p))$$
such that $\lambda_{\perp}(p,v) = \lambda_{\perp}^k(p)$ *for* $v \in V^k(p) \setminus V^{k-1}(p)$.

Proof. By (3.2), $\Pi_{TN_n} \circ d_pf^n \circ \Pi_{TN_0}(v) \equiv d_pf^n \circ \Pi_{TN_0}(v)$. The first statement of (1) then follows (with the proviso $\lambda_{\|} = -\infty$ if $v \in \ker \Pi_{TN_0}$) because definitions (3.6) and (3.3) coincide: note however that, by definition of induced metric in a Riemannian submanifold, $\|.\|_{T_pM} \equiv \|.\|_{T_pN}$ if $v \in T_pN$. (2) follows since the projection $\Pi_{TN_0} : TM_0 \to TN_0$ is surjective.

The second statement in (1), as well as those in (3), (4), (5), are proved in the following way. If $v \in \ker \Pi_{(TN_0)^{\perp}}$ then $\lambda_{\perp}(p,v)$ exists trivially. Otherwise, set $w = \Pi_{(TN_0)^{\perp}} v$. Define a linear map $L_p : (TN_0)^{\perp} \to (TN_1)^{\perp}$ by

$$L_p(w) = \left(\Pi_{(TN_1)^{\perp}} \circ d_pf\right)(w), \qquad w \in (TN_0)^{\perp}$$

and set

$$L_p^n(w) = L_{f^{n-1}(p)} \circ L_{f^{n-2}(p)} \circ \ldots \circ L_p(w).$$

By Lemma 3.2.1 we have

$$L_p^n(w) = \left(\Pi_{(TN_n)^{\perp}} \circ d_pf^n\right)(w).$$

By Oseledec's multiplicative ergodic theorem the limit

$$\lim_{n\to\infty} \frac{1}{n} \log \|L_p^n(w)\|$$

exists for μ-a.a. $p \in A$ and all $w \neq 0$ in $(TN_0)^{\perp}$ (or equivalently all $v \notin \ker \Pi_{TN_0}$ in TM_0), and it takes one of the values $\lambda_{\perp}^1(p) < \ldots < \lambda_{\perp}^{s(p)}(p)$. Statements (3), (4) and (5) are now immediate consequences of Oseledec's theorem. □

If we take μ ergodic in Theorem 3.2.3, the invariant functions $\{\lambda_\perp^i(p), s(p)\}$ are μ-a.e. constant. The normal Liapunov exponents for $v \notin \ker \Pi_{(TN_0)^\perp} \equiv TN_0$ can only take the $s \leq m-n$ values $\lambda_\perp^1 < \ldots < \lambda_\perp^s$ which are independent of p (and the corresponding filtrations invariant under d_pf). In this case we denote these normal Liapunov exponents by $\lambda_\perp^1(\mu) < \ldots < \lambda_\perp^s(\mu)$.

3.2.2 The normal Liapunov spectrum

Definition 3.2.4 *The* measurable normal Liapunov spectrum $S_n(A)$ of A *is*

$$S_n(A) = \bigcup_{\mu \in Erg_f(A)} \{\lambda_\perp^i(\mu)\},$$

where $Erg_f(A)$ is the set of all ergodic invariant probability measures supported in A.

Remark 3.2.5 We call the spectrum 'measurable' because in general the limit in the definition of Liapunov exponents exists only in a set $B \subset A$ such that $\mu(B) = 1$ for all invariant measures $\mu \in \mathcal{M}_f(A)$. In a general chaotic attractor this set may be topologically small; see however Remark 3.2.10. □

From now on we refer to $S_n(A)$ as the normal spectrum, and to

$$\Pi_{(TN_1)^\perp} \circ d_pf \circ \Pi_{(TN_0)^\perp} \stackrel{\text{def}}{=} d_p^\perp f$$

at a point $p \in N$ as the *normal derivative of f at p.*

Lemma 3.2.6 *The normal spectrum $S_n(A)$ is bounded above.*

Proof. For all $p \in A$, all $v \in TM_0$, Lemma 3.2.1 implies that

$$\begin{aligned} \frac{1}{n}\log \|d_p^\perp f^n(v)\| &= \frac{1}{n}\log \|d_{f^{n-1}(p)}^\perp f \circ \ldots \circ d_p^\perp f(v)\| \\ &\leq \frac{1}{n}\log \|d_{f^{n-1}(p)}^\perp f\| \cdot \ldots \cdot \|d_p^\perp f\| \cdot \|v\| \\ &\leq \frac{1}{n}\log(\gamma^n \|v\|) \end{aligned}$$

where $\gamma = \max_{p\in A} \|d_p^\perp f\|$. Therefore for all $p \in A$ and $v \in TM_0$,

$$\limsup_{n\to\infty} \frac{1}{n}\log \|\Pi_{(TN_n)^\perp} d_pf^n(v)\| \leq \gamma.$$

In particular, the normal spectrum $S_n(A)$ is bounded above by γ. □

Remark 3.2.7 If f is non-invertible then the spectrum is necessarily bounded above by Lemma 3.2.6, even though this is not in general an optimal estimate. However, if there are critical points the normal derivative may be singular; in this case, the spectrum need not be bounded below. For instance, suppose A has a fixed point p at which the normal derivative is singular; then $\lambda_{\delta_p} = -\infty$. However, it is also possible that even if there are critical points for the normal derivative the spectrum may be bounded below.

Theorem 3.2.8 below shows that non-singularity of the normal derivative is sufficient for the spectrum to be bounded below. This condition is automatically satisfied, for instance, if f is a diffeomorphism. It also characterizes in an optimal way the extrema of the spectrum. □

Theorem 3.2.8 *Suppose that for all $p \in A$ the normal derivative is injective. Then there exist real $\lambda_{\min} \leq \Lambda_{\max}$ such that:*

1. $S_n(A) \subset [\lambda_{\min}, \Lambda_{\max}]$ *with* $\lambda_{\min}, \Lambda_{\max} \in S_n(A)$*;*
2. $$\inf_{0 \neq v \in (TN_0)^\perp} \liminf_{n\to\infty} \frac{1}{n} \log \|\Pi_{(TN_n)^\perp} d_p f^n(v)\| = \min_{\mu \in Erg_f(A)} \lambda_\perp^1(\mu) = \lambda_{\min},$$
$$\sup_{0 \neq v \in (TN_0)^\perp} \limsup_{n\to\infty} \frac{1}{n} \log \|\Pi_{(TN_n)^\perp} d_p f^n(v)\| = \max_{\mu \in Erg_f(A)} \lambda_\perp^s(\mu) = \Lambda_{\max}.$$

Remark 3.2.9 Note that the projection $\Pi_{(TN_0)^\perp}$ is omitted in the statement and proof of Theorem 3.2.8 since $v \in (TN_0)^\perp$ throughout. □

Proof. Given the normal bundle $NA = \{(p, v) \in TM : p \in A, v \in (TN_0)^\perp\}$, consider the *sphere bundle* $SA = \{(p, v) \in NA : \|v\| = 1\}$. Define the induced tangent map $\widehat{Tf} : SA \to SA$ by

$$\widehat{Tf}(p, v) = \left(f(p), \frac{\Pi_{(TN_1)^\perp} \circ d_p f(v)}{\|\Pi_{(TN_1)^\perp} \circ d_p f(v)\|} \right). \tag{3.8}$$

The requirement of an injective normal derivative ensures that this map is well-defined for all $(p, v) \in SA$, since the denominator in (3.8) never vanishes.

Define $\varphi : SA \to \mathbb{R}$ by

$$\varphi(p, v) = \log \|\Pi_{(TN_1)^\perp} \circ d_p f(v)\|. \tag{3.9}$$

Then:

$$\sum_{i=0}^{n-1} \varphi\left((\widehat{Tf})^i (p, v) \right) = \log \|\Pi_{(TN_1)^\perp} \circ d_p f(v)\| +$$
$$+ \log \frac{\|\Pi_{(TN_2)^\perp} \circ d_p f^2(v)\|}{\|\Pi_{(TN_1)^\perp} \circ d_p f(v)\|} + \ldots + \log \frac{\|\Pi_{(TN_n)^\perp} \circ d_p f^n(v)\|}{\|\Pi_{(TN_{n-1})^\perp} \circ d_p f^{n-1}(v)\|}$$

which gives

$$\sum_{i=0}^{n-1} \varphi\left((\widehat{Tf})^i(p,v)\right) = \log \|\Pi_{(TN_n)^\perp} \circ d_p f^n(v)\|. \tag{3.10}$$

Therefore

$$\limsup_{n\to\infty} \frac{1}{n} \log \|\Pi_{(TN_n)^\perp} \circ d_p f^n(v)\| = \limsup_{n\to\infty} \frac{1}{n} \sum_{i=0}^{n-1} \varphi\left((\widehat{Tf})^i(p,v)\right). \tag{3.11}$$

The right-hand side of (3.11) is a Birkhoff sum. Given (p,v), choose a subsequence $\{n_k\}_{k\geq 0}$ such that

$$\limsup_{n\to\infty} \frac{1}{n} \sum_{i=0}^{n-1} \varphi\left((\widehat{Tf})^i(p,v)\right) = \lim_{k\to\infty} \frac{1}{n_k} \sum_{i=0}^{n_k-1} \varphi\left((\widehat{Tf})^i(p,v)\right).$$

Since SA is compact, the space of probability measures $\mathcal{M}(SA)$ is compact as well. The sequence of probability measures $m_k = \frac{1}{n_k}\sum_{i=0}^{n_k} \delta_{(\widehat{Tf})^i(p,v)}$ therefore has a subsequence converging to a $\widehat{Tf}$-invariant measure $m \in \mathcal{M}_{\widehat{Tf}}(SA)$; therefore $\limsup_{n\to\infty} \frac{1}{n}\sum_{i=0}^{n-1} \varphi\left((\widehat{Tf})^i(p,v)\right) = \int_{SA} \varphi dm$. Thus

$$\sup_{v\in(TN_0)^\perp} \limsup_{n\to\infty} \frac{1}{n} \log \|\Pi_{(TN_n)^\perp} \circ d_p f^n(v)\| \leq \sup_{m\in\mathcal{M}_{\widehat{Tf}}(SA)} \int_{SA} \varphi\, dm. \tag{3.12}$$

Since the converse inequality follows from Oseledec's theorem, (3.12) is in fact an equality. Continuity of φ and compactness of $\mathcal{M}_{\widehat{Tf}}(SA)$ ensure that the sup on the right hand side of (3.12) is attained by some invariant measure; ergodic decomposition (see e.g. Pollicott [95]) and a basic integration property imply that this must happen for an ergodic measure. Thus we may replace the right hand side of (3.12) with

$$\max_{m\in Erg_{\widehat{Tf}}(SA)} \int_{SA} \varphi\, dm.$$

All that remains is to relate the integrals $\int_{SA} \varphi dm$ to the Liapunov exponents in A. To do this, note that in the commutative diagram

$$\begin{array}{ccc} SA & \xrightarrow{\widehat{Tf}} & SA \\ \pi_1 \downarrow & & \downarrow \pi_1 \\ A & \xrightarrow{f} & A \end{array} \tag{3.13}$$

where π_1 is projection onto the first factor, an ergodic $\widehat{Tf}$-invariant measure m projects via $\mu = m \circ \pi_1^{-1}$ to an ergodic f-invariant measure on A. Part (5) of

Theorem 3.2.3 and injectivity of the normal derivative imply that there is a set $A_\mu \subset A$ of full μ-measure such that, if $p \in A_\mu$ and $V^i(p)$ is the subspace at p associated with the normal Liapunov exponent $\lambda^i_\perp(\mu)$, then

$$\Pi_{(TN_1)^\perp} \circ d_p f(V^i(p)) = V^i(f(p)). \tag{3.14}$$

Define $E^i = \bigcup_{p\in A_\mu} V^i(p)$, and $\widehat{E^i}$ to be their restrictions to SA. Then the $\widehat{E^i}$ satisfy the following:

$$\widehat{E^1} \subset \widehat{E^2} \subset \ldots \subset \widehat{E^s} = SA, \tag{3.15}$$

$$\widehat{Tf}(\widehat{E^i}) = \widehat{E^i}, \tag{3.16}$$

both equations being valid m-a.e.. Since $m(\widehat{E^s}) = m(SA) = 1$ it follows by ergodicity that m must be concentrated on some $\widehat{E^i}$, that is, $m(\widehat{E^i}) = 1$ and $m(\widehat{E^{i-1}}) = 0$ for some $1 \le i \le s$. Then

$$\int_{SA} \varphi \, dm = \int_{SA} \log \|\Pi_{(TN_1)^\perp} \circ d_p f(v)\| \, dm$$

must be a constant m-a.e. But Oseledec's theorem and definition of the $\widehat{E^i}$ imply that for m-a.a. $(p, v) \in \widehat{E^i}$

$$\lambda^i(\mu) = \lim_{n\to\infty} \frac{1}{n} \sum_{k=0}^{n-1} \varphi(\widehat{Tf}^k(p, v)) = \int_{SA} \varphi \, dm,$$

so that this constant must equal $\lambda^i(\mu)$. This and (3.12) show that

$$\sup_{v\in(TN_0)^\perp} \limsup_{n\to\infty} \frac{1}{n} \log \|\Pi_{(TN_n)^\perp} \circ d_p f^n(v)\| = \max_{\mu\in Erg_f(A)} \{\lambda^s_\perp(\mu)\}.$$

A similar argument shows that

$$\inf_{v\in(TN_0)^\perp} \liminf_{n\to\infty} \frac{1}{n} \log \|\Pi_{(TN_n)^\perp} \circ d_p f^n(v)\| = \min_{\mu\in Erg_f(A)} \{\lambda^1_\perp(\mu)\}.$$

Injectivity of the normal derivative and compactness of A imply that

$$\inf_{\|v\|=1} \|\Pi_{(TN_1)^\perp} \circ d_p f(v)\|$$

is uniformly bounded away from zero, so an argument analogous to that on Lemma 3.2.6 shows that the normal spectrum is contained in a compact interval. □

Remark 3.2.10 If there exists a point $p \in A$ with a dense orbit such the normal Liapunov exponents do not exist, i.e. if the set of limit points of

$$\frac{1}{n} \log \|d_p f^n(v)\| \tag{3.17}$$

is non-trivial for some $v \in TN_0$, then it may be shown that normal Liapunov exponents only exist in a set of first Baire category in A; see [27]. However, Theorem 3.2.8 implies that for all $p \in A$ and all non-zero $v \in TM_0$ the set of limit points of (3.17) is always contained in $[\Lambda_{\min}, \Lambda_{\max}]$. Therefore the spectrum bounds the possible asymptotic growth rates in the normal direction. □

3.2.3 Stability indices

Given an ergodic invariant probability measure $\mu \in Erg_f(A)$, the normal Liapunov exponents $\lambda_\perp^1(\mu) < \ldots < \lambda_\perp^s(\mu)$ exist and are constant in a set B_μ of full μ-measure. The normal stability or instability of this set is determined by the sign of the largest normal Liapunov exponent. We therefore associate to every ergodic μ its *normal stability index* Λ_μ in the following way:

Definition 3.2.11 *Let μ be an f-invariant ergodic probability measure supported in A, with normal Liapunov exponents $\lambda_\perp^1(\mu) < \ldots < \lambda_\perp^s(\mu)$. The* normal stability index *Λ_μ of μ is*

$$\Lambda_\mu = \lambda_\perp^s(\mu).$$

Note that $\Lambda_{\max} = \max_{\mu \in Erg_f(A)}\{\Lambda_\mu\}$. The definition is useful because of the following result.

Theorem 3.2.12 *Suppose A is an asymptotically stable attractor for $f_{|N}$. Then:*

1. *If $\Lambda_{\max} < 0$ then A is an asymptotically stable attractor for f in M.*
2. *If $\Lambda_{\max} > 0$ then A is Liapunov-unstable.*

The proof of Theorem 3.2.12 is divided into several steps. Firstly, by Lemma 1.3.9 we may take a compact neighbourhood $W \subset N$ of A in N in the relative topology such that $f(W) \subset W$ and $\bigcap_{n=0}^{\infty} f^n(W) = A$. Define the *normal bundle* of W to be

$$TW^\perp = \bigcup_{p \in W} (T_pN)^\perp.$$

An element of $TW^\perp$ is thus (p, v) where $p \in W$, $v \in (T_pN)^\perp$.

Consider the function $g : TW^\perp \to M$ given by

$$g(p, v) = \exp_p(v), \tag{3.18}$$

where $\exp : TM \to M$ is the exponential map. By compactness of W, there exists $\delta > 0$ such that $g(p, v)$ is defined for all $(p, v) \in TW^\perp$ if $\|v\| < \delta$. In other words, g is defined in a neighbourhood of the zero-section of $TW^\perp$.

Lemma 3.2.13 *There exists $\epsilon > 0$ such that g is a diffeomorphism between*

$$\tilde{W}_\epsilon \stackrel{\text{def}}{=} \{(p, v) \in TW^\perp : \|v\| \leq \epsilon\}$$

and a compact neighbourhood of A in M.

Remark 3.2.14 By a standard property of the exponential map, $g(p, 0)$ is the identity – that is, for all $(p, 0) \in TN^\perp$, $g(p, 0) = p \in M$. □

Proof. First we prove that $g : TW^\perp \to M$ is a local diffeomorphism at a point $(p, 0)$ of the zero-section of $TW^\perp$. We construct a basis of $T_{(p,0)}(TW^\perp)$ and T_pM respecting the splittings $T_{(p,0)}(TW^\perp) \cong T_pN \times (T_pN)^\perp$, $T_pM \cong T_pN \times (T_pN)^\perp$. Since $g(p, v) = \exp_p(v)$, $v \in (T_pN)^\perp$, and taking into account that $\exp_p(0) \equiv p$, $\frac{d}{dt}\left(\exp_p(tv)\right)\big|_{t=0} = v$, we conclude that the derivative $d_{(p,0)}g$ is given in such a basis by the matrix

$$d_{(p,0)}g = \begin{pmatrix} \mathbf{1} & 0 \\ 0 & \mathbf{1} \end{pmatrix}.$$

By the inverse function theorem, g is a local diffeomorphism between neighbourhoods of $(p, 0) \in TW^\perp$ and of $p \in M$. By compactness of W there exists $\eta > 0$ such that for all $p \in W$, g is a local diffeomorphism on the ball $B(p, v)$ with $\|v\| < \eta$.

Consider the compact set $\tilde{W}_\epsilon = \bigcup_{p \in W}\{(p, v) : \|v\| \leq \epsilon\}$ for $\epsilon < \eta/2$. We claim that there exists $\epsilon > 0$ such that $g_{|\tilde{W}_\epsilon}$ is a diffeomorphism from $\tilde{W}_\epsilon$ onto its image.

If not, we can find sequences $y_n \in M$, $p_n^{(1)}, p_n^{(2)} \in W$, $v_n^{(1)}, v_n^{(2)} \in (TN_0)^\perp$ such that

$$y_n = \exp_{p_n^{(1)}}(v_n^{(1)}) = \exp_{p_n^{(2)}}(v_n^{(2)}) \tag{3.19}$$

and $d(y_n, W) \to 0$. By compactness of W and relabelling, we may suppose all these sequences converge. Then $y_n \to p \in N$. Therefore, for large enough n, $(p_n^{(1)}, v_n^{(1)})$ and $(p_n^{(2)}, v_n^{(2)})$ must lie in the η-ball around $(p, 0)$, since $\exp_p(0) = p$. But then equality (3.19) contradicts the fact that g is a local diffeomorphism on this ball. We conclude that there exists $\epsilon > 0$ such that g is injective in $\tilde{W}_\epsilon$, and therefore $g_{|\tilde{W}_\epsilon}$ is a diffeomorphism onto its image, as claimed. □

Denoting the image of $\tilde{W}_\epsilon$ under g by W_ϵ, we conclude that there is a $\delta > 0$ such that, defining

$$\tilde{f} = g^{-1} \circ f \circ g, \tag{3.20}$$

the diagram

$$\begin{array}{ccc} \tilde{W}_\epsilon & \xrightarrow{\tilde{f}} & \tilde{W}_\delta \\ g \downarrow & & \downarrow g \\ W_\epsilon & \xrightarrow{f} & W_\delta \end{array} \tag{3.21}$$

commutes.

Proof of Theorem 3.2.12. Taking the derivative of (3.20) at $(p,0)$ gives

$$\begin{aligned} d\tilde{f}_{(p,0)} &= dg^{-1} \circ df \circ dg_{(p,0)} \\ &= dg^{-1}_{f(p)} \circ d_p f \\ &= d_p f \end{aligned}$$

where the obvious identifications of tangent spaces are made. Therefore, by induction

$$\Pi_{(TN_n)^\perp} \circ d_{(p,0)}\tilde{f}^n \circ \Pi_{(TN_0)^\perp} = \Pi_{(TN_n)^\perp} \circ d_p f^n \circ \Pi_{(TN_0)^\perp}, \tag{3.22}$$

that is, f and $\tilde{f}$ have the same normal derivatives. As $g(p,0)$ is the identity, the normal Liapunov exponents of a point $p \in A$ under f and $\tilde{f}$ coincide. In particular, the normal spectra $S_n(A)$ of A under f and $\tilde{f}$ are equal.

From now on, denote by

$$d_p^\perp \tilde{f} = \Pi_{(TN_1)^\perp} \circ d_{(p,0)}\tilde{f} \circ \Pi_{(TN_0)^\perp}$$

the normal derivative of $\tilde{f}$ at $(p,0) \in W$. Let $\epsilon > 0$ and $\delta > 0$ be such g is a diffeomorphism and diagram (3.21) commutes. For $0 < \alpha < \min\{\epsilon, \delta\}$, take the compact neighbourhood $\tilde{W}_\alpha$ in $TW^\perp$ defined by

$$\tilde{W}_\alpha = \{(p,v) \in TW^\perp : p \in W,\ \|v\| \le \alpha\}. \tag{3.23}$$

Fix λ such that $\Lambda_{\max} < \lambda < 0$. By compactness of W, continuity of $d_p^\perp \tilde{f}^n$ and Theorem 3.2.8 we conclude that there exists n_0 such that $n \ge n_0$ implies, for all $p \in W$, all $v \in TN_0$ with $v \ne 0$,

$$\|d_p^\perp \tilde{f}^n(v)\| < e^{n\lambda}\|v\|.$$

Choose n such that $e^{(n-1)\lambda} < \frac{1}{3}$. Denote by

$$(p_n, v_n) = \tilde{f}^n(p,v).$$

Thus $p_n = \pi_1 \tilde{f}^n(p,v)$, $v_n = \pi_2 \tilde{f}^n(p,v)$, where π_1, π_2 are the projections on the first and second factors of $TW^\perp$ respectively.

Our choice of $W \subset N$ implies that $\tilde{f}(W) \subset W$, where inclusion is strict. Set $a = d_H(\tilde{f}^n(W), W) > 0$, where d_H is the Hausdorff distance between compact sets in $TW^\perp$. We may take $\alpha > 0$ sufficiently small in (3.23) that

$$\rho(\tilde{f}^n(p,v), \tilde{f}^n(p,0)) < a/3 \tag{3.24}$$

for all $(p,v) \in \tilde{W}_\alpha$, where ρ denotes the Riemannian metric induced on $TW^\perp$ by that of M through the diffeomorphism g, and

$$\rho(\tilde{f}^n(p,v), \tilde{f}(p,0)) < e^{-\lambda/2}\|d_p^\perp \tilde{f}^n(v)\|. \tag{3.25}$$

Condition (3.24) may be met by uniform continuity of $\tilde{f}^n$, and condition (3.25) by continuity of $d_p^\perp f$.

With these choices,

$$\begin{aligned}\|v_n\| &\leq \rho(\tilde{f}^n(p,v),\tilde{f}^n(p,0))\\ &< e^{-\lambda/2}\|d_p^\perp \tilde{f}^n(v)\|\\ &< e^{-\lambda/2}e^{n\lambda}\|v\|\\ &< e^{(n-1)\lambda}\|v\|\\ &< \frac{1}{3}\|v\|. \end{aligned}\tag{3.26}$$

Equation (3.24) implies that $\tilde{f}^n(\tilde{W}_\alpha) \subset V_{\alpha/2}(\tilde{f}^n(W))$, where $V_\delta(\tilde{f}^n(W))$ is the δ-neighbourhood of $\tilde{f}^n(W)$. In particular this implies $p_n = \pi_1 \tilde{f}^n(p,v) \in W$. These conditions together imply that $(p_n, v_n) = \tilde{f}^n(p,v) \in \tilde{W}_\alpha$ for all $(p,v) \in \tilde{W}_\alpha$, or $\tilde{f}^n(\tilde{W}_\alpha) \subset \tilde{W}_\alpha$. Setting $K = \bigcup_{j=0}^{n-1} \tilde{f}^j(\tilde{W}_\alpha)$, we conclude that K is compact and forward-invariant. By equation (3.26) we see that for all $(p,v) \in K$, $\|v_n\| \to 0$, which implies that

$$\bigcap_{n\geq 0} \tilde{f}^n(K) \subset \bigcap_{n\geq 0} \tilde{f}^n(W) = A,$$

where the last equality is due to asymptotic stability of A as a subset of N. Therefore A is an asymptotically stable attractor for the map $\tilde{f}$.

Translating these results via the diffeomorphism g into M, we conclude that A is an asymptotically stable attractor for the map $f: M \to M$.

Proof of part 2: Suppose that $\Lambda_{\max} > 0$. Then there exists an ergodic invariant measure μ supported in A with $\Lambda_\mu > 0$ by Theorem 3.2.8. Fix $0 < \lambda < \Lambda_\mu$. By Theorem 3.2.3, for μ-a.a. $p \in A$ there exists a filtration

$$\{0\} = V^0 \subset V^1 \ldots \subset V^l = (TN_0)^\perp$$

of $(TN_0)^\perp$ with

$$\lim_{n\to\infty} \frac{1}{n} \log \|\Pi_{(TN_n)^\perp} \circ d_p f^n(v)\| = \Lambda_\mu$$

for $v \in V^l \setminus V^{l-1}$.

As in part 1, we consider the commutative diagram (3.21), the diffeomorphism g and equation (3.22). Define

$$M_n = \{(p,0) \in W_\epsilon : m \geq n \Rightarrow \frac{1}{m} \log \|d_p^\perp \tilde{f}^m(v)\| > \lambda,\ \|v\| = 1, v \in V^l \setminus V^{l-1}\}.$$

By definition $\mu(M_n) \to 1$ as $n \to \infty$; we take n large enough that $M_n \neq \emptyset$. Therefore, if $p \in M_n$,

$$\|d_p^\perp \tilde{f}^m(v)\| > e^{m\lambda}\|v\|$$

for $v \in V^l(p) \setminus V^{l-1}(p)$, all $m \geq n$.

Using the same notation as in part (1), choose $n > 0$ such that

$$\|d_p^\perp \tilde{f}^n(v)\| > e^{n\lambda}\|v\|$$

with $e^{(n-1)\lambda} > 2$ if $v \in V^l \setminus V^{l-1}$. Choose α small enough that

$$\|v_n\| > e^{-\lambda/2}\|d_p^\perp \tilde{f}^n(v)\|$$

for all $(p, v) \in \tilde{W}_\alpha$. Then

$$\|v_n\| > e^{-\lambda/2}e^{n\lambda}\|v\| > 2\|v\|.$$

Thus, for any α, $\bigcup_{n=0}^\infty \tilde{f}^{-n}(\tilde{W}_\alpha)$ cannot contain a neighbourhood of A. As the $\{\tilde{W}_\alpha\}_{\alpha>0}$ form a basis of neighbourhoods of W in $TW^\perp$, this implies that there is no neighbourhood V of A such that $\bigcup_{n=0}^\infty \tilde{f}^{-n}(V)$ contains a neighbourhood of A – in other words, A is Liapunov unstable in $TW^\perp$. Applying the diffeomorphism g to W_ϵ, we conclude that A is Liapunov unstable in M for the original map f. □

We next give a sufficient condition for the basin of a Milnor attractor to be locally riddled.

Given an ergodic measure μ, we denote the set of (measure-theoretical) *generic points* of μ (see Denker *et al.* [27]) by G_μ. That is,

$$G_\mu = \{x \in A : \frac{1}{n}\sum_{j=0}^{n-1} \delta_{f^j(x)} \to \mu\}$$

where convergence is in the weak* topology. For any $\alpha > 0$ define

$$G_\alpha = \bigcup_{\substack{\mu \in Erg_f(A) \\ \Lambda_\mu \geq \alpha}} G_\mu.$$

Theorem 3.2.15 *Suppose A is a Milnor attractor for f and $\Lambda_{\max} > 0$. If there exists $\alpha > 0$ such that G_α is dense in A, then A has a locally riddled basin.*

Proof. We set $\beta = \alpha/2$ and argue by contradiction. Suppose A does not have a locally riddled basin. Setting $U(V) = \bigcap_{n=0}^\infty f^{-n}(V)$ as in Definition 3.1.2, this means that for every neighbourhood V of A there exist $p \in A$ and $\delta > 0$ such that $\ell(B_\delta(p) \cap U(V)^c) = 0$. As we want to consider arbitrarily small compact neighbourhoods of A in M, we may restrict attention to those contained inside neighbourhoods where g as in Lemma 3.2.13 and in diagram (3.21) is a diffeomorphism. Zero measure sets, local basin riddling and density of G_α are clearly invariant under the diffeomorphism g^{-1}, so the previous statements translate automatically to the corresponding ones in $\tilde{W}_\epsilon$. Moreover if $p \in N$ then $g^{-1}(p) = (p, 0) \in TW^\perp$ and α in the statement remains invariant under g.

Given $0 < \gamma < \epsilon$, define as before

$$\tilde{W}_\gamma = \{(p, v) \in TW^\perp : p \in W, \|v\| \leq \gamma\},$$

where W is a small enough compact invariant neighbourhood of A in N. Then the $\tilde{W}_\gamma$ are compact and form a neighbourhood basis for W in $TW^\perp$.

For convenience we work below with the product metric in $TW^\perp$: for $p \in N$, we set

$$B_\delta(p) = \{(q, v) \in TW^\perp : \max(\rho_N(q, p), \|v\|) < \delta\},$$

where ρ_N is the Riemannian distance induced on N. We follow the notations of Theorem 3.2.12: $d_p^\perp \tilde{f} \equiv \Pi_{(TN_1)^\perp} \circ d_{(p,0)}\tilde{f} \circ \Pi_{(TN_0)^\perp}$ is the normal derivative of f at $p \in A$, $(p_n, v_n) = \tilde{f}^n(p, v)$ and π_i is the projection onto the ith factor of $(p, v) \in TW^\perp$, $i = 1, 2$.

Choose $\tilde{W}_\gamma$ sufficiently small that

$$\|v_1\| > e^{-\beta/4} \|d_p^\perp \tilde{f}\| \; \|v\| \tag{3.27}$$

and

$$\|d_{p_1}^\perp \tilde{f}\| > e^{-\beta/4} \|d_{\tilde{f}(p)}^\perp \tilde{f}\| \tag{3.28}$$

for all (p, v) in $\tilde{W}_\gamma$. This is possible by continuity of $d_p^\perp \tilde{f}$ and compactness of $\tilde{W}_\gamma$. Note that $p_1 = \pi_1 \tilde{f}(p, v) \neq \tilde{f}(p, 0)$ in general, as the non-linear map $\tilde{f}$ does not preserve the fibres over W.

Suppose $p_0 \in A$ and $\delta > 0$ are such that $B_\delta(p_0) \cap U(\tilde{W}_\gamma)^c$ has zero measure. By density of G_α there exists $q \in G_\alpha \cap B_\delta(p_0)$. Then there exists n_0 such that

$$n \geq n_0 \Rightarrow 1/n \log \|d_q^\perp \tilde{f}^n\| > \beta,$$

or equivalently

$$\|d_q^\perp \tilde{f}^n\| > e^{n\beta}.$$

Consider the smallest n satisfying $e^{\frac{n\beta}{2}} > \gamma/\delta$. By continuity of $d_p^\perp \tilde{f}^n$ there exists a compact disk $D_\eta(q) \subset B_\delta(p_0) \cap N$ such that

$$p \in D_\eta(q) \Rightarrow \|d_p^\perp \tilde{f}^n\| > e^{n\beta}. \tag{3.29}$$

As the $d_p^\perp \tilde{f}^n$ are finite-dimensional linear operators their norms must be attained by vectors $v(p)$; moreover these vary continuously with p. Thus for each $p \in D_\eta(q)$

$$\|d_p^\perp \tilde{f}^n(v(p))\| > e^{n\beta} \|v(p)\|. \tag{3.30}$$

Consider the vertical strip $S_{\eta,\delta} = D_\eta(q) \times \{v : \|v\| \leq \delta\} \subset B_\delta(p_0)$. Then, for $(p, v) \in S_{\eta,\delta}$,

$$\begin{aligned}
\|v_n\| &> e^{-\beta/4}\|d^\perp_{p_{n-1}}\tilde{f}\|\|v_{n-1}\| \\
&> e^{-2\beta/4}\|d^\perp_{p_{n-1}}\tilde{f}\|\|d^\perp_{p_{n-2}}\tilde{f}\|\|v_{n-2}\| \\
&\vdots \\
&> e^{-n\beta/4}\|d^\perp_{p_{n-1}}\tilde{f}\|\|d^\perp_{p_{n-2}}\tilde{f}\|\cdots\|d^\perp_{p_1}\tilde{f}\|\|d^\perp_{p}\tilde{f}\|\|v(p)\|
\end{aligned}$$

where inductive use is made of (3.27). We now apply inductively (3.28):

$$\begin{aligned}
\|v_n\| &> e^{-(n+1)\beta/4}\|d^\perp_{p_{n-1}}\tilde{f}\|\|d^\perp_{p_{n-2}}\tilde{f}\|\cdots\|d^\perp_{p_2}\tilde{f}\|\|d^\perp_{\tilde{f}(p)}\tilde{f}\|\|d^\perp_{p}\tilde{f}\|\|v(p)\| \\
&\vdots \\
&> e^{-n\beta/2}\|d^\perp_{\tilde{f}^{n-1}(p)}\tilde{f}\|\|d^\perp_{\tilde{f}^{n-2}(p)}\tilde{f}\|\cdots\|d^\perp_{\tilde{f}(p)}\tilde{f}\|\|d^\perp_{p}\tilde{f}\|\|v(p)\| \\
&\geq e^{-n\beta/2}\|d^\perp_{p}\tilde{f}^n\|\|v(p)\| \\
&> e^{n\beta/2}\|v(p)\| \\
&> \frac{\gamma}{\delta}\|v(p)\|.
\end{aligned}$$

Taking $v(p)$ with $\|v(p)\| = \delta$ for each p, we have $\|\pi_2 \circ \tilde{f}^n(q, v(q))\| > \gamma$. Continuity of $v(p)$ in p and non-singularity of $d^\perp\tilde{f}^n$ imply that $\tilde{f}^n(S_{\eta,\delta}) \cap \tilde{W}_\gamma^{\,c}$ contains an open set, and has therefore positive measure, contradicting the hypotheses. Therefore A has a locally riddled basin in $TW^\perp$. This statement translates through the diffeomorphism g to the corresponding one in M, proving that A has a locally riddled basin in M. □

3.2.4 Chaotic saddles

In this section we consider the situation diametrically opposed to that of Theorem 3.2.12. We assume that the normal spectrum lies completely to the right of the origin and conclude, using essentially the same methods, that A is normally repelling: the only points which remain in a small neighbourhood of A for all iterations are those already lying in the submanifold N. In particular, if A is a chaotic attractor for $f_{|N}$, then A is a normally repelling chaotic saddle.

Theorem 3.2.16 *Suppose $\lambda_{\min} > 0$. Then for any sufficiently small neighbourhood U of A, $f^n(p) \in U$ for all $n \geq 0$ if and only if $p \in N$.*

Proof. The if part is trivial since A is an asymptotically stable attractor for $f_{|N}$. We concentrate on the only if part.

The notation in this proof will be that of Theorems 3.2.12 and 3.2.15. As before, the statement in the Theorem translates through g^{-1} to $TW^\perp$.

Choose λ such that $0 > \lambda > \lambda_{\min}$. By Theorem 3.2.8, there exists n such that, for all $p \in W$, $m \geq n$, $0 \neq v \in (TN_0)^\perp$,

$$\|d_p^\perp \tilde{f}^m(v)\| > e^{m\lambda}\|v\|. \tag{3.31}$$

Choose a small enough compact neighbourhood $\tilde{W}_\gamma$ of A in $TW^\perp$ that $\tilde{f}^n(\tilde{W}_\gamma)$ remains inside the neighbourhood referred to in Lemma 3.2.13 and

$$\|\pi_2 \circ \tilde{f}^n(p,v)\| > e^{-\lambda/2}\|d_p^\perp \tilde{f}^n(v)\|. \tag{3.32}$$

Let $U = \{(p,v) \in V : \tilde{f}^j(p,v) \in V \text{ for all } j \geq 0\} = \bigcap_{j=0}^\infty \tilde{f}^{-j}(V)$. Notice that the zero-section of $(TW)^\perp$ is contained in U since by hypothesis A is an asymptotically stable attractor for $f_{|N}$. Moreover, U is forward-invariant: $\tilde{f}(U) \subseteq U$.

If $(p,v) \in U$, then by definition $\tilde{f}^j(p,v) \in V$ for all $j \geq 0$. But by (3.31) and (3.32)

$$\|\pi_2 \tilde{f}^n(p,v)\| > e^{-\lambda/2}e^{n\lambda}\|v\| \tag{3.33}$$

for all $(p,v) \in V$. Equation (3.33) shows that if $\|v\| \neq 0$ there exists k such that $\tilde{f}^{kn}(p,v) \notin V$; thus $(p,v) \notin U$. Therefore U coincides with the zero-section W_0 of $W \cap N$ in $(TW)^\perp$.

Applying the diffeomorphism g this translates to f in M as the 'only if' part in the statement of this Theorem. □

3.2.5 Strong SBR measures

By assuming existence of a strong SBR measure we can say more about the normal stability of A. Let $f : M \to M$ be a $C^{1+\alpha}$ map; as before, suppose that M has an invariant embedded submanifold N such that A is an asymptotically stable attractor for $f_{|N}$. Suppose in addition that A has a strong SBR measure under $f_{|N}$ as in Definition 1.4.16, which we denote by μ_{SBR}, with k positive and $n-k$ negative Liapunov exponents. As A is an asymptotically stable attractor for f, it must be the closure of the union of unstable manifolds on N. The unstable manifolds W_σ induce a measurable partition of A, see Pugh and Shub [97]. Denote this partition by

$$A = \bigcup_{\sigma \in \Sigma} W_\sigma,$$

each W_σ being an open piece of a k-dimensional unstable manifold that carries a conditional measure μ_σ such that, for any Borel measurable set $B \subset A$,

$$\mu_{SBR}(B) = \int \mu_\sigma(B \cap W_\sigma)\, d\mu_\sigma.$$

It follows from work of Pesin [89], Ledrappier and Strelcyn [61], Pugh and Shub [97], that the existence of a measure which is absolutely continuous on unstable manifolds ensures the *absolute continuity of the stable foliation of* A. A family of stable manifolds has this property if for any family of local stable disks D^s_α of dimension $n-k$ in M and any pair of smooth manifolds V_1, V_2 of dimension k which intersect the D^s_α transversely, $\ell_{V_i}(V_i \cap \bigcup_\alpha D^s_\alpha)$ are both zero or both strictly positive, $i = 1, 2$, where ℓ_{V_i} denotes the Riemannian measure on V_i induced by the Riemannian structure on M.

We now state a powerful result of Alexander *et al.* [1].

Theorem 3.2.17 *Let A be a strong SBR attractor for $f_{|N}$. Suppose $\Lambda_{SBR} < 0$. Then $\ell_M(\mathcal{B}(A)) > 0$. Moreover, if A is either uniformly hyperbolic or μ_{SBR} is absolutely continuous with respect to an n-dimensional Riemannian measure on N then A is an essentially asymptotically stable attractor.*

Proof. See Alexander *et al.* [1]. □

This statement is a consequence of Theorem 3.2.12 in the case where $\Lambda_{\max} < 0$, since A is then an asymptotically stable attractor for f in the global phase space M. The new case covered by Theorem 3.2.17 is when $\Lambda_{\max} > 0 > \Lambda_{SBR}$. By Theorem 3.2.12, if $\Lambda_{\max}$ is positive then A cannot be asymptotically stable – indeed, it is Liapunov unstable. This may be pictured in terms of the normal unstable manifolds of A. Let μ be an ergodic measure with $\Lambda_\mu > 0$. Then by a result of Ruelle [102] for μ-a.a. $x \in A$ there exist local unstable manifolds $W^u_{loc}(x)$ not tangent to N, corresponding to the positive normal Liapunov exponents. Thus A is not the closure of the unstable manifolds of its points; these manifolds "stick out" away from N and so they intersect any neighbourhood of A in M, making A Liapunov unstable. However, the condition $\Lambda_{SBR} < 0$ implies that ℓ_N-a.a. $x \in A$ have empty normal unstable manifolds, and therefore A attracts a set of positive Lebesgue measure. Under the stronger assumptions in the Theorem, the measure of this set relative to that of a neighbourhood of A tends to 1 as the neighbourhood shrinks. Therefore A is an essentially asymptotically stable attractor.

As the examples discussed in § 3.4 and in [1], [82] show, this is the best one can in general hope for due to the presence of repelling 'tongues' of positive measure in any neighbourhood of a normally unstable ergodic subset of such an attractor.

The proof of Theorem 3.2.17 in [1] may be adapted to yield the following converse. We conjecture that the hypotheses for the following theorem are typically satisfied, but do not have a general proof of this.

Theorem 3.2.18 *Suppose the attractor A of $f_{|N}$ is a nontrivial SBR attractor. Suppose $\Lambda_{SBR} > 0$, and furthermore all Liapunov exponents are μ_{SBR}-a.e. different from zero and that $\ell_M(\cup_{\mu \neq \mu_{SBR}} G_\mu) = 0$. Then $\ell_M(\mathcal{B}(A)) = 0$; that is, A is a chaotic saddle.*

Proof. Suppose that A has μ_{SBR}-a.e. $k \geq 1$ positive and $n-k$ negative tangent Liapunov exponents: this is because $\Lambda_{SBR} > 0$, implying A has at least one normal Liapunov exponent μ_{SBR}-a.e. greater than zero. By ergodicity of μ_{SBR}, there are exactly $m-n$ normal exponents, including multiplicity. Suppose that of these $m-n$ normal exponents d $(\neq 0)$ are positive and $m-n-d$ are negative (by assumption they are all nonzero).

Since μ_{SBR} is absolutely continuous on unstable manifolds in N, we can choose a local unstable manifold $W_\sigma \subset N$ of A such that the conditional measure μ_σ is absolutely continuous with respect to a Riemannian measure on W_σ.

Denote by W'_σ the local unstable manifold of A, of dimension $k+d$, such that

$$W_\sigma = W'_\sigma \cap N.$$

By absolute continuity, μ_σ-a.a. $p \in W_\sigma$ has a local stable manifold D^s_p of dimension $m-(k+d)$ which is transverse to W'_σ. The assumption that non SBR-generic points of the attractor have zero measure implies that we need only consider the stable disks at points that are generic for μ_{SBR}. Furthermore, by transversality $W'_\sigma \cap D^s_p$ is a single point, namely p. However, we also have $W_\sigma \cap D^s_p = \{p\}$, so that for any μ_σ-measurable $B \subset W_\sigma$,

$$W'_\sigma \cap \bigcup_{p \in B} D^s_p = W_\sigma \cap \bigcup_{p \in B} D^s_p \subset N, \tag{3.34}$$

so that this intersection is, for all $B \subset W_\sigma$, a subset of the k-dimensional unstable manifold W_σ. Hence, denoting by ℓ'_σ the Riemannian measure induced on the $(k+d)$-dimensional manifold W'_σ, we conclude that

$$\ell'_\sigma\left(W'_\sigma \cap \bigcup_{p \in B} D^s_p\right) = 0. \tag{3.35}$$

As discussed in [1], for sufficiently small $\delta > 0$ we can find $B_\delta \subset W_\sigma$ with positive ($W_\sigma-$) Riemannian measure such that the local stable manifold D^s_p makes an angle at least δ with W'_σ and extends at least a distance δ in all $m-(k+d)$ directions with curvature less than $1/\delta$. Moreover, $\mu_\sigma(B_\delta) \to 1$ as $\delta \to 0$.

Next we smoothly foliate a δ^2-tubular neighbourhood $V_{\delta,\sigma}$ about W'_σ with smooth $(k+d)$-dimensional manifolds V_β transverse to $\bigcup_{p \in B_\delta} D^s_p$. Absolute continuity of the stable foliation is a consequence of the fact that all $m-n$ normal exponents are different from zero and of Theorem 2 in Pugh and Shub [97]. Together with equation (3.35) this guarantees that

$$\ell_{V_\beta}\left(V_\beta \cap \bigcup_{p \in B_\delta} D^s_p\right) = 0,$$

and consequently the Riemannian measure of $\bigcup_{p \in B_\delta} D^s_p$ is zero. □

Remark 3.2.19 On comparing the statements of Theorems 3.2.17 and 3.2.18, we see that their conclusions are dramatically different. Theorem 3.2.17 states that, if the SBR measure is normally stable, the basin of A has positive measure, even though it may be riddled with a dense set of "holes" of positive measure – as is the case of the mentioned "repelling tongues". Theorem 3.2.18 deals with the case where the SBR measure in normally unstable; it states that the basin of A has zero measure, even though it may contain an uncountable family of stable manifolds. In a short and descriptive language, this result states that "attractive tongues", analogous to the repelling tongues in the opposite setting, do not exist.

3.2.6 Classification by normal spectrum

For convenience, we collect the results of the previous sections together in the following proposition.

Proposition 3.2.20 *Suppose $f : M \to M$ is a $C^{1+\alpha}$ map leaving the embedded submanifold N invariant, and that A is an asymptotically stable chaotic attractor for $f_{|N}$. Define $\Lambda_{\max}$, $\lambda_{\min}$ and Λ_{SBR} (if there is an SBR measure supported on A) as above. Then, under $f : M \to M$:*

(a) *If $\Lambda_{\max} < 0$ then A is an asymptotically stable attractor.*
(b) *If $\Lambda_{\max} > 0$ then A is Liapunov unstable.*
(c) *If $\Lambda_{SBR} < 0 < \Lambda_{\max}$ then A is a Milnor attractor. If in addition there exists $\alpha > 0$ with G_α dense in A, then A has a locally riddled basin.*
(d) *If $\lambda_{\min} < 0 < \Lambda_{SBR}$, μ_{SBR}-almost all Liapunov exponents are non-zero and $\ell_M(\cup_{\mu \neq \mu_{SBR}} G_\mu) = 0$ then A is a chaotic saddle.*
(e) *If $0 < \lambda_{\min}$ then A is a normally repelling chaotic saddle.*

Proof. This is a collection of the results in Theorems 3.2.12, 3.2.15, 3.2.16, 3.2.17, and 3.2.18. □

Remark 3.2.21 (Normal Hyperbolicity, perturbations breaking N) Normal hyperbolicity guarantees persistence of invariant manifolds under small perturbations of the dynamical system [86]. We remark that this study is taking place precisely in the region where $A \subset N$ is not normally hyperbolic. In particular, if A is normally hyperbolic, a riddled basin cannot exist; conversely, if A has a riddled basin then small arbitrary perturbations of f will break up the manifold N. For experimental systems, this has the implication that small departures from the conditions that create N – e.g. noise or a small asymmetry – may destroy the attractor altogether, leaving behind what has been termed a "ghost", see Venkataramani *et al.* [117].

3.3 Normal parameters and normal stability

The previous section is a 'static' classification of normal dynamics using the normal spectrum; until now we have not discussed parameters. The parameter dependence of a chaotic attractor is very delicate for non-uniformly hyperbolic attractors. Rather than considering this question, we concentrate on studying how the transverse behaviour of the attractor changes when the dynamics on the invariant submanifold N are fixed. To do this, we define a *normal parameter* of the system – one that preserves the dynamics on the invariant submanifold but varies it in the rest of the phase space.

Even when the normal dynamics is continuously dependent on a normal parameter, it does not follow that the normal spectrum varies continuously. However, for systems that satisfy a certain technical condition (they map a continuous cone field in the tangent bundle into itself) it is possible to prove continuity of the normal spectrum using a result of Ruelle [101]. Since this condition is open in the space of maps on the tangent bundle, we can expect to observe it.

Using this assumption, we prove in Theorem 3.3.3 that there is an open set in an appropriate function space in which the normal spectrum is continuously dependent on normal parameters. We can therefore meaningfully discuss the transitions between an asymptotically stable attractor, a locally riddled basin attractor, a chaotic saddle and a normally repelling chaotic saddle upon variation of normal parameters. In particular if N has codimension 1 in M the cone condition is automatically satisfied and these bifurcations occur generically.

Theorem 3.3.5 suggests that we can expect to observe attractors with locally riddled basins in a persistent way. We characterize a generic loss of stability in Proposition 3.3.9. Liapunov exponents only give a linearized theory of stability; in order to calculate branching behaviour at bifurcations, global stable and unstable manifolds must be investigated. We discuss this and aspects of the dimensions of bifurcating attractors in § 3.3.2.

3.3.1 Parameter dependence of the normal spectrum

We consider the set of maps of M that are equivalent to f as in § 3.2.1 when restricted to a neighbourhood of A in the invariant submanifold N. To this end, consider a compact neighbourhood $U \subset M$ of A.

Definition 3.3.1 *Given $h \in C^1(N,N)$ we say $f \in C^1(U,M)$ is an* extension *of h if $f_{|U\cap N} = h_{|U\cap N}$. We define the* set of extensions *of h in U to be $\mathcal{F}_U(h)$.*

For discussion of the parameter dependence it is useful to consider paths in $\mathcal{F}_U(h)$ for fixed h and U. In fact, allowing the perturbations to h to be in a general function space would have the effect that the invariant attractor A for h would vary discontinuously (upper-semicontinuously at best) and the whole structure of $Erg_f(A)$ would vary discontinuously; this problem is outside the scope of the present work.

Definition 3.3.2 *We say ν is a* normal parameter *if there is an $h \in C^1(N,N)$ such that $f(\cdot,\nu) \in \mathcal{F}_U(h)$ is smoothly dependent on ν.*

This means that a normal parameter does not affect the dynamics on N. Consequently, normal parameters preserve all invariant measures on $A \subset N$, and so $Erg_{f(\cdot,\nu)}(A)$ is independent of ν. As we shall see in § 3.5, normal parameters appear naturally if the invariant subspace N is forced by coupling between identical systems. An important observation is that tangent Liapunov exponents are independent of normal parameters. This follows directly from formula (3.6).

Theorem 3.3.3 *Suppose $h \in C^1(N, N)$ has an asymptotically stable attractor A. Then there exists a non-empty open subset $\tilde{\mathcal{F}}_U(h)$ of $\mathcal{F}_U(h)$ such that, for all $\mu \in Erg_f(A)$, the normal Liapunov exponents $\lambda_\perp^i(\mu, \nu)$ are smooth functions of ν.*

Before proving Theorem 3.3.3 we need some definitions. Recall $\dim N = n$, $\dim M = m$ and define

$$\mathcal{T} = C^0(U \cap N, GL(\mathbb{R}^{m-n})),$$

the Banach space of continuous maps from $U \cap N$ to the space of $(m-n) \times (m-n)$ matrices equipped with the supremum norm (recall U is compact).

Define a compact neighbourhood $W \subset U \cap N$ of A, a smooth splitting $TW = \{(p, T_pN \oplus (T_pN)^\perp) \ : \ p \in W\}$ and a diffeomorphism $g : (TW)^\perp \to W_\epsilon$ as in § 3.2.1. We define $\tilde{f} = g^{-1} \circ f \circ g$.

For this splitting, there is a natural restriction map $\mathcal{R} : \mathcal{F}_U(h) \to \mathcal{T}$ given by

$$(\mathcal{R}f)(p) = \Pi_{(T_{h(p)}N)^\perp} \circ d_{(p,0)}f \circ \Pi_{(T_pN)^\perp} \equiv d_p^\perp f$$

for $p \in N$, where $\mathbb{R}^{m-n}$ is equated with $(T_pN)^\perp$. This map is well defined. It is also surjective; given $(p, v) \in (TW)^\perp$ we define

$$\tilde{f}(p, v) = (h(p), M(p)v)$$

and note that $\mathcal{R}f = M$ (this map is a skew product that preserves the foliation). $\mathcal{R}$ is continuous with respect to the topologies in $\mathcal{F}_U(h)$ and $\mathcal{T}$. Note that the normal Liapunov exponents of $f \in \mathcal{F}_U(h)$ are exactly the Liapunov exponents of the matrix product defined by $\mathcal{R}f$.

Proof of theorem 3.3.3. By a result of Ruelle [101], there exists a non-empty open set $\tilde{\mathcal{T}} \subset \mathcal{T}$ (corresponding to linear maps that map a continuous cone bundle into itself) such that the Liapunov exponents are analytic in $\tilde{\mathcal{T}}$. Continuity of $\mathcal{R}$ implies that its preimage is an open set $\tilde{\mathcal{F}}(h)$. Since $f(.,\nu)$ is assumed smooth as a function of ν, the normal Liapunov exponents $\lambda_\perp^i(\mu, \nu)$ will be smooth in ν. □

Remark 3.3.4 If codim $N = 1$, there is only one normal direction. In this case $\lambda_\mu = \Lambda_\mu$ for all ergodic μ, and there is a unique cone field as in [101] which trivially satisfies the invariance condition. In this case the normal spectrum depends smoothly on normal parameters, i.e. $\tilde{\mathcal{F}}_U(h) = \mathcal{F}_U(h)$. In fact we can in principle compute the normal exponents explicitly; see (3.43) in § 3.4 below. However, for codim $N > 1$ the inclusion $\tilde{\mathcal{F}}_U(h) \subset \mathcal{F}_U(h)$ is strict. □

We now show that $\lambda_{\min} < \Lambda_{SBR} < \Lambda_{\max}$ holds generically in $\tilde{\mathcal{F}}_U(h)$ in the C^1 topology if A is a non-trivial (i.e. not a fixed or periodic point) Axiom A attractor.

Theorem 3.3.5 *Suppose $h \in C^{1+\alpha}(N, N)$ has a non-trivial Axiom A attractor. Then, generically in $\tilde{\mathcal{F}}_U(h)$, Λ_{SBR} is not an extremum of the normal spectrum.*

Proof. We show that $\Lambda_{SBR} < \Lambda_{\max}$ generically; the proof that $\lambda_{\min} < \Lambda_{SBR}$ is similar. Continuity of Λ_μ with $f \in \tilde{\mathcal{F}}_U(h)$ for any $\mu \in Erg_h(A)$ implies that $\Lambda_{SBR} < \Lambda_{\max}$ is satisfied in an open subset of $\tilde{\mathcal{F}}_U$.

To prove density, choose $f \in \tilde{\mathcal{F}}_U(h)$ such that $\Lambda_{SBR} = \Lambda_{\max}$. As A satisfies Axiom A, periodic point measures are dense in $\mathcal{M}_f(A)$; see Sygmund [116]. Let x be a periodic point such that

$$\Lambda_\mu(f) > \Lambda_{SBR}(f) - \frac{\epsilon}{3} \tag{3.36}$$

where μ is the ergodic measure supported on the orbit $\mathcal{O}(x)$ of x.

Define W, $\tilde{W}_\gamma$, $\tilde{f}$ and g as in the proof of Theorem 3.2.12. Let $\tilde{\varphi} = g^{-1} \circ \varphi \circ g$ and recall that the normal Liapunov exponents at $p \in A$ are identical for f and $\tilde{f}$. Define $\tilde{\psi} : \tilde{W}_\gamma \to \tilde{W}_\gamma$ by

$$\tilde{\psi}(p, v) = (p, e^{\epsilon\eta(p,v)}v)$$

so that

$$d_{(p,v)}\tilde{\psi} = \begin{pmatrix} \mathbf{1} & \epsilon\frac{\partial\eta}{\partial p}e^{\epsilon\eta}v \\ 0 & e^{\epsilon\eta(p,v)}\mathbf{1} + \epsilon\frac{\partial\eta}{\partial v}e^{\epsilon\eta}v \end{pmatrix}$$

For any $\delta > 0$ with $B_\delta(p, 0) \subset \tilde{W}_\gamma$ in the product metric, define $\eta : \tilde{W}_\gamma \to \mathbb{R}$ to be

$$\eta(p, v) = \zeta(\min_{q\in\mathcal{O}(x)}(\rho_N(p, q))/\delta)\zeta(\|v\|/\delta)$$

where ρ_N is the Riemannian distance in N and $\zeta : [0, \infty) \to [0, 1]$ is given by

$$\zeta(a) = \begin{cases} 1 & \text{for } 0 \leq a \leq 1 \\ \frac{1}{2}(1 + \cos(a\pi - \pi)) & \text{for } 1 \leq a \leq 2 \\ 0 & \text{for } a \geq 2 \end{cases}$$

Note that ζ, η and $\tilde{\psi}$ are continuously differentiable and

$$\left\|\frac{\partial\eta}{\partial p}\right\|_0 + \left\|\frac{\partial\eta}{\partial v}\right\|_0 < \frac{4}{\delta}.$$

Define $\tilde{\varphi} : \tilde{W}_\epsilon \to \tilde{W}_\epsilon$ by

$$\tilde{\varphi}(p, v) = \tilde{\psi} \circ \tilde{f}(p, v). \tag{3.37}$$

For any $p \in A$, $0 \neq v \in (T_pN)^\perp$, the normal Liapunov exponents $l_\perp(p, v)$ at p for $\tilde{\varphi}$ are:

$$l_\perp(p, v) = \lim_{n\to\infty}\frac{1}{n}\log\|\Pi_{(TN_n)^\perp}d_{(p,0)}\tilde{\varphi}^n(v)\|_{TM_n}.$$

Since

$$d_{(p,0)}\tilde{\psi} = \begin{pmatrix} \mathbf{1} & 0 \\ 0 & e^{\epsilon\eta(p,0)}\mathbf{1} \end{pmatrix}$$

it follows that

$$\begin{aligned} l_\perp(p,v) &= \lim_{n\to\infty}\frac{1}{n}\log\left(\|d_p^\perp \tilde{f}^n(v)\|\prod_{k=0}^{n-1} e^{\epsilon\eta(f^k(p),0)}\right) \\ &= \lim_{n\to\infty}\frac{1}{n}\left(\log\|d_p^\perp \tilde{f}^n(v)\| + \sum_{k=0}^{n-1}\epsilon\eta(f^k(p),0)\right) \\ &= \lambda_\perp(p,v) + \epsilon\lim_{n\to\infty}\frac{1}{n}\sum_{k=0}^{n-1}\eta(f^k(p),0), \end{aligned}$$

where $\lambda_\perp(p,v)$ are the normal Liapunov exponents at p for f. For $m \in Erg_f(A)$ it follows from Theorem 3.2.8 that $\lambda_\perp$ exists and is constant on a set of full m-measure. On this set the Ergodic Theorem implies

$$l_\perp(p,v) = \lambda_\perp(p,v) + \epsilon\int \eta(p,0)\,dm$$

for all $v \in (T_pN)^\perp$. In particular, the independence of the integral term on v implies that $\Lambda_m(\tilde{\varphi}) = \Lambda_m(\tilde{f}) + \epsilon\int\eta(p,0)\,dm$. Thus

$$\Lambda_\mu(\varphi) = \Lambda_\mu(f) + \epsilon \tag{3.38}$$

and

$$\Lambda_{SBR}(\varphi) = \Lambda_{SBR}(f) + \epsilon\int\eta(p,0)\,d\mu_{SBR}.$$

Since μ_{SBR} is non-trivial we can choose δ such that $\int\eta(p,0)\,d\mu_{SBR} < \frac{1}{3}$; thus

$$\Lambda_{SBR}(\varphi) < \Lambda_{SBR}(f) + \frac{\epsilon}{3}. \tag{3.39}$$

Combining equations (3.36), (3.38) and (3.39) we obtain

$$\frac{\epsilon}{3} + \Lambda_{SBR}(\varphi) < \Lambda_\mu(\varphi) \leq \Lambda_{\max}(\varphi). \tag{3.40}$$

Choosing $\epsilon > 0$ such that $e^\epsilon - 1 < 2\epsilon$ (and $e^\epsilon < 2$) we have

$$\begin{aligned} \|d_{(p,v)}\tilde{\psi} - \mathbf{1}\|_0 &< \|e^{\epsilon\eta} - 1\|_0 + \epsilon\left\|\frac{\partial\eta}{\partial p}e^{\epsilon\eta}v\right\|_0 + \epsilon\left\|\frac{\partial\eta}{\partial v}e^{\epsilon\eta}v\right\|_0 \\ &< 2\epsilon + 2.\epsilon.\frac{4}{\delta}.2.2\delta \\ &= 34\epsilon, \end{aligned}$$

and thus

$$\|d_{(p,v)}\tilde{\varphi} - d_{(p,v)}\tilde{f}\|_0 < 34\epsilon\|d_{(p,v)}\tilde{f}\|_0 < 34\epsilon\|\tilde{f}\|_1.$$

From (3.37) we have also $\|\tilde{\varphi} - \tilde{f}\|_0 < 2\epsilon\|\tilde{f}\|_1$; therefore

$$\|\tilde{\varphi} - \tilde{f}\|_1 < 36\epsilon \max\{\|\tilde{f}\|_1, 1\}.$$

Thus $\tilde{\varphi}$ is arbitrarily C^1-close to $\tilde{f}$ and $\Lambda_{SBR}(\tilde{\varphi}) < \Lambda_{\max}(\tilde{f})$. This statement translates to M as the density part of the Theorem. □

Remark 3.3.6 This result is probably true under weaker hypotheses. In particular, we expect that genericity corresponds to codimension infinity in the sense that even in generic k-parameter families in the $C^{1+\alpha}$ topology, we do not expect to see Λ_{SBR} as an extremum of the spectrum. □

Remark 3.3.7 If there exists a dense set of periodic points or a set of periodic points with a dense set of preimages, we expect to observe local riddling of basins. This is because the condition

$$\Lambda_{SBR} < 0 < \Lambda_{\max}$$

is satisfied generically in $\tilde{\mathcal{F}}_U(h)$; then G_α should generically become dense in A prior to Λ_{SBR} crossing zero and we apply Theorem 3.2.15. □

We are now in a position to discuss the bifurcation behaviour of A.

Definition 3.3.8 *Let $\nu \in \mathbb{R}$ be a normal parameter for f. We define the following bifurcation points for ν.*

- ν_0 *is a* point of loss of asymptotic stability *if $\Lambda_{\max}(\nu)$ passes through zero at $\nu = \nu_0$.*
- ν_0 *is a* blowout bifurcation point [83] *if $\Lambda_{SBR}(\nu)$ passes through zero at $\nu = \nu_0$.*
- ν_0 *is a* point of bifurcation to normal repulsion *if $\lambda_{\min}(\nu)$ passes through zero at $\nu = \nu_0$.*

Note that if on varying a normal parameter A changes from being an asymptotically stable attractor to a normally repelling chaotic saddle, we are forced in a persistent way (that is, generically in an open subset of $\mathcal{F}_U(h)$) to undergo a sequence of bifurcations:

Proposition 3.3.9 *Suppose that $f(\cdot,\nu)$ is a smooth path in $\tilde{\mathcal{F}}_U(h)$ and that h has an asymptotically stable attractor A. If for $f(\nu_0)$, the set A is asymptotically stable and for $f(\nu_1)$ it is a normally repelling chaotic saddle then generically there exist $\nu_0 < \nu_a < \nu_b < \nu_c < \nu_1$ such that:*

- *There is a loss of asymptotic stability at ν_a.*
- *There is a blowout bifurcation at ν_b.*
- *There is a bifurcation to normal repulsion at ν_c.*

Proof. This follows from noting that $\Lambda_{\max}(\nu_0) < 0$, $\lambda_{\min}(\nu_1) > 0$ and using continuity of $\Lambda_{\max}(\nu)$ and $\lambda_{\min}(\nu)$ from Theorems 3.3.3 and 3.3.5. □

3.3.2 Global behaviour near bifurcations

In Definition 3.3.8 we characterized the important bifurcations of the normal stability of A in terms of the Liapunov exponents. The branching behaviour near such bifurcations is not determined at linear order. Whether the bifurcations are determined locally or globally depends on the type of invariant measure that becomes unstable.

Bifurcations from a periodic orbit $\{p, f(p), \dots, f^{n-1}(p)\}$ with $f^n(p) = p$ and $p \in A$ can be dealt with by considering the nth iterate of the map. Each point in the orbit is a fixed point for f^n, and the bifurcations from it are generically determined by quadratic or higher order terms of the Taylor expansion of f at the fixed point. In this sense, the branching behaviour is determined locally.

For a blowout bifurcation of the SBR measure on A, any branching from A will be considerably complicated by the presence of a dense set of unstable manifolds even before bifurcation. We discuss some more speculative aspects of the blowout bifurcation in § 3.6, motivated by the numerical experiments in § 3.4.

3.4 Example: $\mathbb{Z}_2$-symmetric maps on $\mathbb{R}^2$

We consider two families of extensions of a cubic logistic equation $h : \mathbb{R} \to \mathbb{R}$ defined by

$$h(x) = \frac{3\sqrt{3}}{2} x(x^2 - 1)$$

to a map of the plane. Let $f_{\alpha,\nu,\epsilon}$ and $g_{\alpha,\nu,\epsilon}$ be three-parameter maps of $\mathbb{R}^2$ to itself that are equivariant under $\mathbb{Z}_2$ generated by $(x, y) \mapsto (x, -y)$, and given by:

$$f_{\alpha,\nu,\epsilon}\begin{pmatrix} x \\ y \end{pmatrix} = \begin{pmatrix} \frac{3\sqrt{3}}{2}x(x^2-1) + \epsilon x^2 y \\ \nu e^{-\alpha x^2} y + y^3 \end{pmatrix} \tag{3.41}$$

$$g_{\alpha,\nu,\epsilon}\begin{pmatrix} x \\ y \end{pmatrix} = \begin{pmatrix} \frac{3\sqrt{3}}{2}x(x^2-1) + \epsilon x^2 y \\ \nu e^{-\alpha x^2 - y^2} y + \frac{(1-e^{-y^2})}{2} y \end{pmatrix}. \tag{3.42}$$

The factor $\frac{3\sqrt{3}}{2}$ is such that each of the intervals $[-1, 0]$ and $[0, 1]$ are mapped onto the other in a two-to-one way (with the obvious exceptions of the critical points and the origin). In fact, h has an asymptotically stable attractor $A = [-1, 1] \subset N$ independently of α, ν and ϵ. All three parameters are normal parameters and these families are in $\tilde{\mathcal{F}}(h)$; their restrictions to $N = \mathbb{R} \times \{0\}$ coincide. Moreover, $d_{(x,0)}f = d_{(x,0)}g$: the derivative maps coincide, and so in particular the normal derivatives are the same. Consequently, the normal spectra of f

and g coincide. The normal derivatives on N are dependent on α and ν but independent of ϵ. The case $\epsilon = 0$ corresponds to the existence of an invariant foliation by vertical lines.

Although these two maps have the same normal spectra, the global dynamics of f and g are quite different. Note that f has a (superstable) attractor at infinity whereas g has a repellor at infinity.

3.4.1 The spectrum of normal Liapunov exponents

For both families,

$$d_{(x,0)}f = d_{(x,0)}g = \begin{pmatrix} 3\sqrt{3}(x^2 - \frac{1}{2}) & \epsilon x^2 \\ 0 & \nu e^{-\alpha x^2} \end{pmatrix}.$$

and $\Lambda_\mu = \lambda_\mu$ since there is only one normal direction. We can explicitly compute this stability index:

$$\begin{aligned} \lambda_\mu = \Lambda_\mu &= \int_A \log \|\Pi_{(T_{h(x,0)}N)^\perp} \circ d_{(x,0)}f \circ \Pi_{(T_{(x,0)}N)^\perp}\| d\mu(x) \\ &= \int_A \log |\nu e^{-\alpha x^2}| \, d\mu(x), \\ &= \log|\nu| - K_\mu \alpha, \end{aligned} \tag{3.43}$$

where we write

$$K_\mu = \int_A x^2 d\mu(x).$$

For all invariant measures, K_μ is finite (indeed $0 \le K_\mu \le 1$); moreover, $\lambda_\mu = \Lambda_\mu$ is smoothly dependent on $\alpha \in \mathbb{R}$. This is apparent from the linearity of (3.43); indeed this fact is a consequence of Remark 3.3.4, since we are dealing with a codimension 1 case and would be true regardless of the specific form of the map. We write K_{SBR} for $K_{\mu_{SBR}}$ and $K_{\max}$ for $\max_{\mu \in Erg_f(A)} K_\mu$. Note that for fixed α, ν, the lower extremum of the spectrum, $\lambda_{\min}$, is by (3.43) equal to $\log|\nu| - K_{\max}\alpha$. Note also that the upper extremum of the spectrum is $\Lambda_{\max} = \log|\nu|$: the origin is a fixed point and the normal derivative there is ν (in other words $K_{\delta_{(0,0)}} = 0$).

It is a standard exercise to show that f has an absolutely continuous ergodic measure μ_{SBR} whose support is A. The coordinate change $x = \sin^3(\pi\theta/2)$ turns $f_{|A}$ into an expanding map; by a result of Lasota and Yorke [60] it follows that f has an ergodic absolutely continuous invariant measure. As explained in §3.2, normal stability of this measure plays a prominent role as it marks the difference between A being a Milnor attractor or a chaotic saddle in the global phase space. Numerical approximation of the SBR measure from box counting 500,000 iterates in 100 bins gives

$$K_{SBR} = 0.358.$$

Numerical evidence also suggests that the invariant measure on A having the largest value of K (and therefore realizing $\lambda_{\min}$) is supported on the symmetric period two point at

$$x = \pm\sqrt{1 - \frac{2}{3\sqrt{3}}}$$

implying that

$$K_{\max} = 1 - \frac{2}{3\sqrt{3}} = 0.615.$$

Figure 3.1 (p. 96 ff.) shows a sequence of pictures of the basin $\mathcal{B}(A)$ in $\mathbb{R}^2$ for $\alpha = 0.7$ and increasing values of ν; Figure 3.2 (p. 98 ff.) shows blow-ups of details from this figure. Figure 3.3 (p. 100) shows approximations of the area $\ell(\mathcal{B}(A))$ upon varying ν.

We now analyse in detail the nature of the invariant set A in the global phase space using the methods developed in § 3.2. For simplicity we state the results only for positive ν; note, however, that this classification is essentially independent of the sign of ν.

3.4.2 Global transverse stability for f

The fact that $\Lambda_{\max} = \log \nu$ coupled with the second part of Theorem 3.2.12 implies that, for $\nu > 1$, A is Liapunov unstable in the global phase space. A direct analysis is enlightening.

Theorem 3.4.1 *For any $\epsilon \in \mathbb{R}$ and $\alpha > 0$, the behaviour of the map f is as follows.*

1. *For $0 \le \nu < 1$, A is an asymptotically stable attractor.*
2. *For $1 < \nu < e^{K_{SBR}\alpha}$, A is a Milnor (essential) attractor whose basin is riddled with that of the attractor at infinity.*
3. *For $e^{K_{SBR}\alpha} < \nu < e^{K_{\max}\alpha}$, A is a (non-normally repelling) chaotic saddle.*
4. *For $\nu > e^{K_{\max}\alpha}$, A is a normally repelling chaotic saddle.*

Proof. (1) It follows from (3.43) that $\Lambda_\mu \le \log \nu$ for all invariant measures μ. The origin is a fixed point of (3.41), so $\delta_{(0,0)}$ is an invariant measure for f_A. Indeed, as $\Lambda_{\delta_{(0,0)}} = \log \nu$ from (3.43), it follows that for all α, ϵ, $\Lambda_{\max} = \log \nu$. Hence for $0 \le \nu < 1$ Theorem 3.2.12 implies that A is an asymptotically stable attractor. This is true both for f and g since, as noted above, their restrictions and linearizations on A coincide.

(2) We know from Theorem 3.2.12 that A is Liapunov unstable. As noted above, when $\nu > 1$ the origin is a hyperbolic source with a two-dimensional unstable manifold. The eigendirections associated to this unstable manifold the coordinate axes. Note that, for $\nu > 1$, f is a local diffeomorphism in a neighbourhood of the origin; it follows from the Hartman-Grobman Theorem [85] that there is a neighbourhood of the origin in which the dynamics of f is topologically conjugate

to those of its linearization. This linearization admits a one-parameter family of codimension 1 local invariant submanifolds given by

$$W_k^{loc} = \{(x,y) \in U : |x| = k|y|^a\}, \tag{3.44}$$

where U is the neighbourhood of the origin in the Hartman-Grobman theorem and $a = \frac{\log \nu}{\log(\frac{3\sqrt{3}}{2})}$. So, for $1 < \nu < \frac{3\sqrt{3}}{2}$ the situation is as depicted in Figure 3.4 (p. 101).

Extending this family of local submanifolds by topological continuation (that is, $W_k = \bigcup_{n\geq 0} f^n(W_k^{loc})$) produces a foliation of the 2-dimensional unstable manifold of the origin by invariant submanifolds.

We next observe that the line $x = 0$ is actually invariant for the dynamics. Moreover, as it contains no fixed points, it follows that W_0 is the whole y-axis, which is in the basin of ∞. It is easy to check that the complement of the strip $|y| \leq 1$ is in the basin of ∞ for all $\nu > 0$; it follows by continuity that there is a neighbourhood of 0 in the variable k such that the submanifolds W_k cross the lines $|y| = 1$ and are thus in the basin of ∞. We thus get the global picture depicted in Figure 3.1(c).

Therefore, for $\nu > 1$ there is a cusp-shaped open set near the origin which is repelled away from A and attracted towards ∞, which we call a “repulsive tongue” T_0; see Remark 3.4.2.

Figure 3.1 (Page 96 ff.)
The basin of attraction $\mathcal{B}(A)$ for $f_{\alpha,\nu,\epsilon}$ as a function of ν with $\alpha = 0.7$, $\epsilon = 0.5$. The black area denotes the (approximate) basin of attraction while the colours indicate the number of iterates requires to escape from the disk $\sqrt{x^2+y^2} < 10$. The region shown is the rectangle $(x,y) \in [-1.5, 1.5] \times [-1.1, 1.1]$.

(a) For $\nu = 0.9$ the attractor is asymptotically stable, though the basin boundary seems to be fractal in nature.

(b) At $\nu = 1.2$ Theorem 3.4.1 implies that the basin is riddled. This is not apparent in the picture since the measure of the ‘holes’ is too small to be detected numerically.

(c) At $\nu = 1.28$, near the blowout bifurcation, there is a complex fractal structure visible in the basin. Figure 3.2 shows details from this basin.

(d) At $\nu = 1.48$ the basin has zero measure, but still extends away from the invariant x-axis; A is a non-normally repelling chaotic saddle.

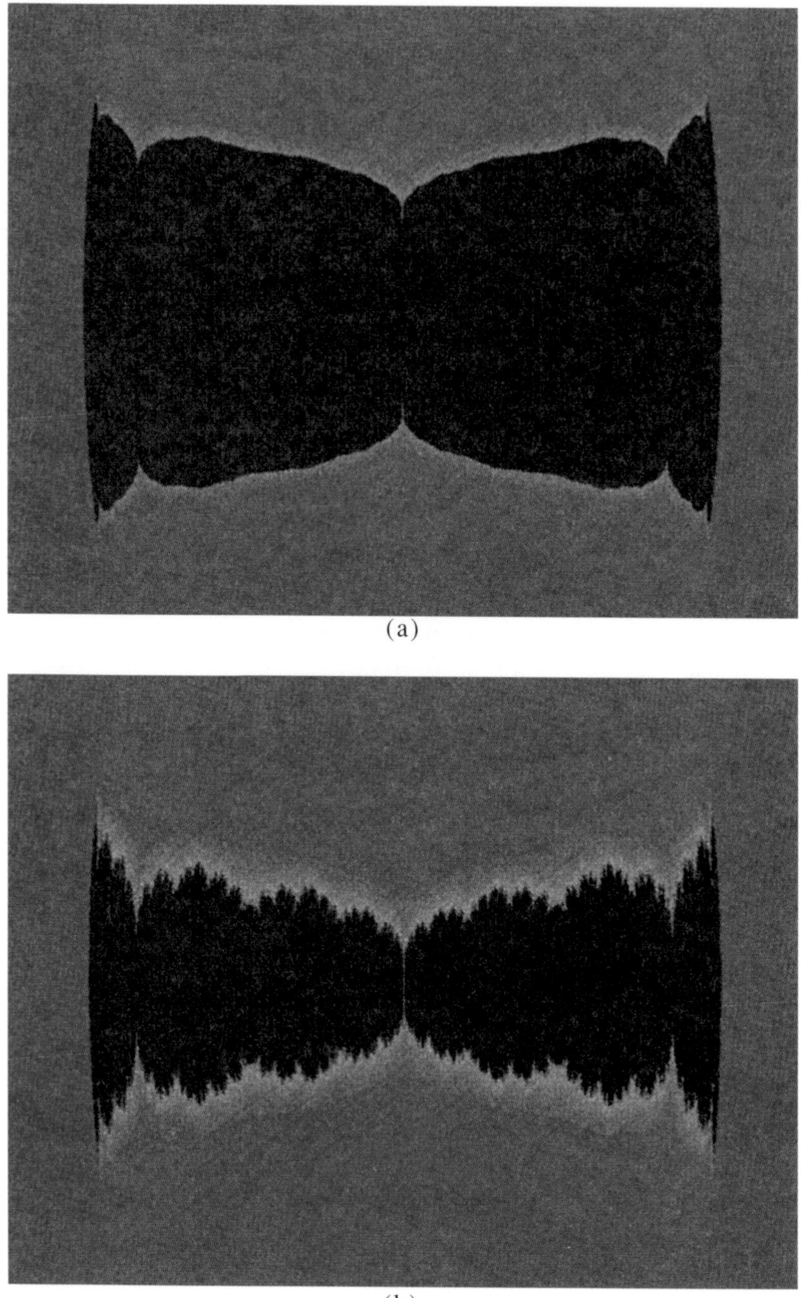

Figure 3.1: Basin of attraction of f as a function of ν

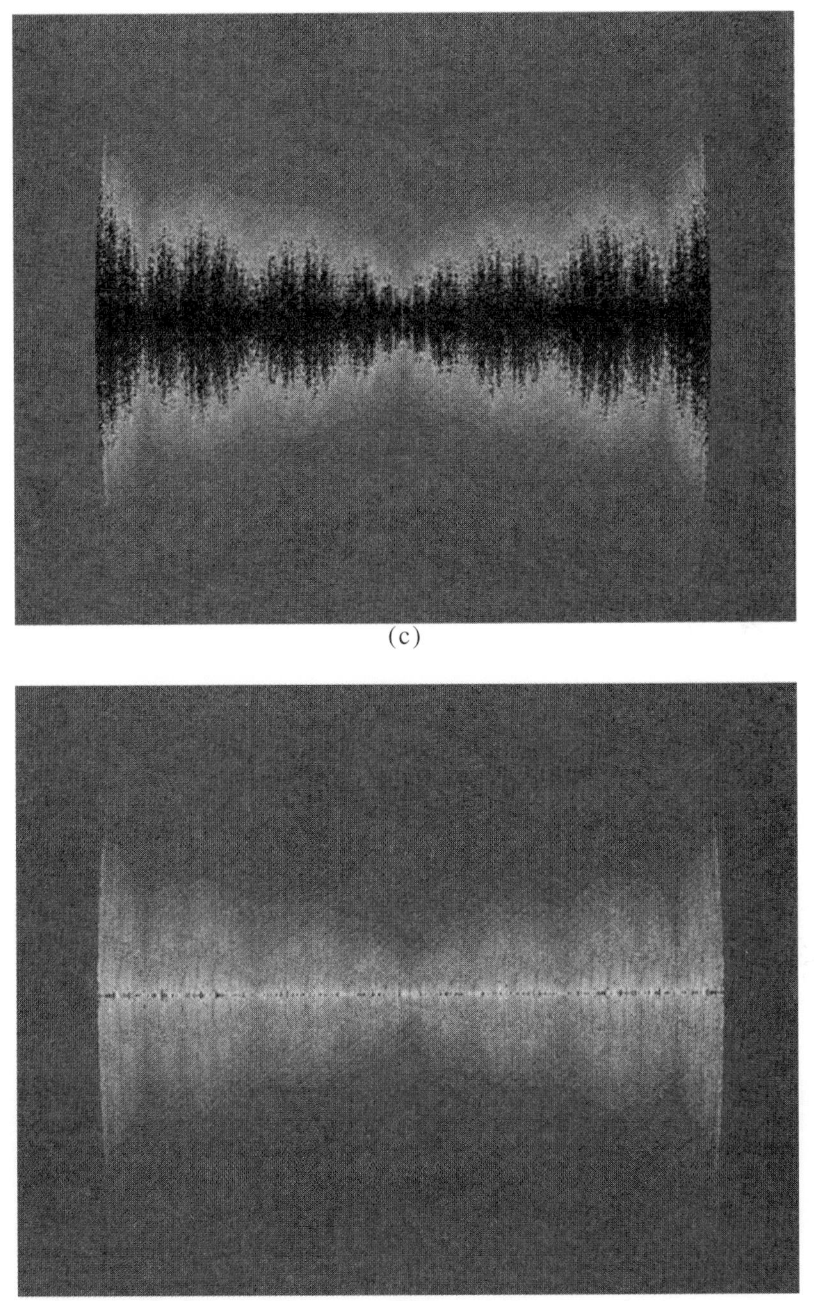

(c)

(d)

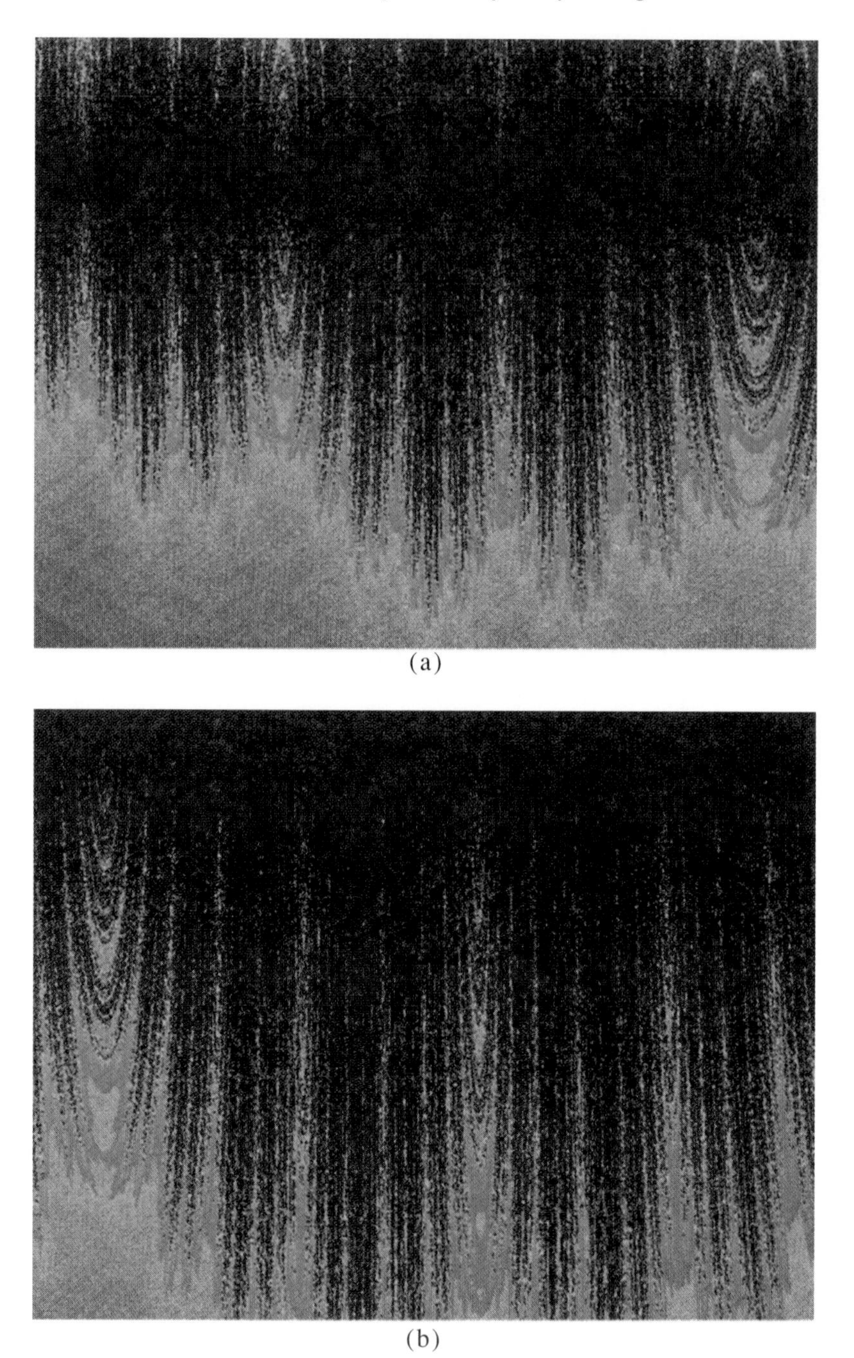

(a)

(b)

Figure 3.2: Zooming in on fig. 3.1(c)

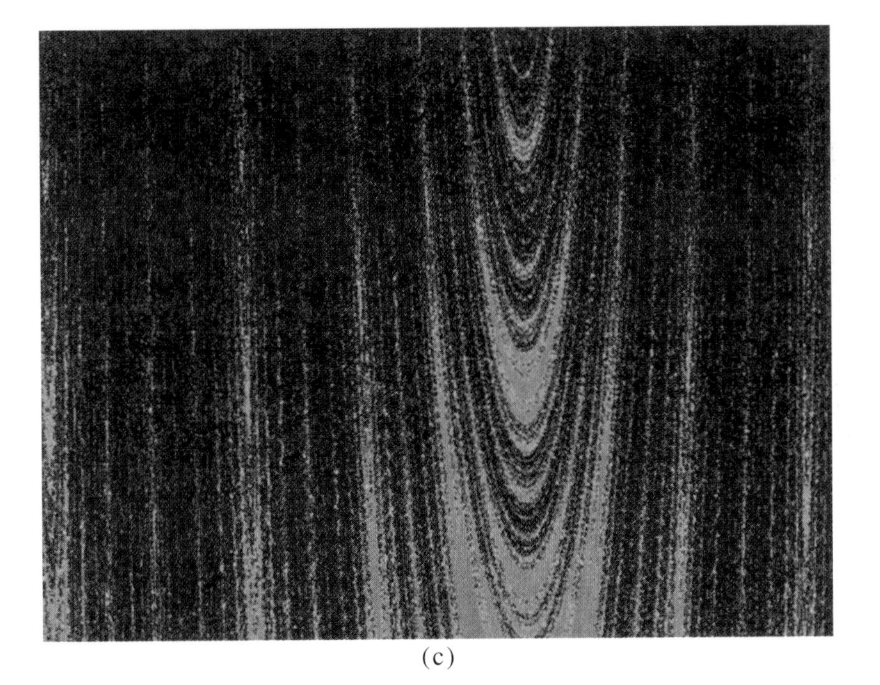

(c)

Figure 3.2 Details of Figure 3.1(c) are shown for (x, y) in the following regions:

(a) $[0.048, 0.623] \times [-0.104, 0.317]$;

(b) $[0.204, 0.306] \times [-0.005, 0.239]$;

(c) $[0.150, 0.250] \times [0.039, 0.134]$.

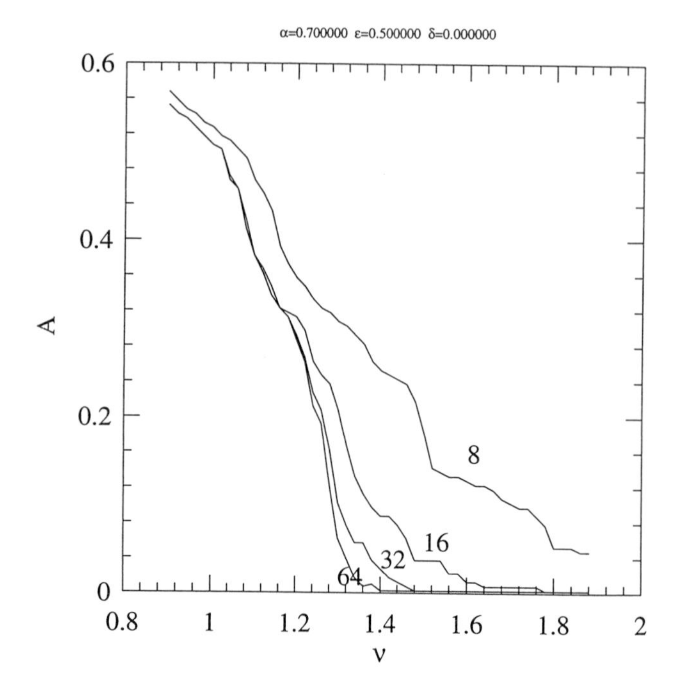

Figure 3.3: Area of the basin of attraction of A as a function of ν. This was found by numerically computing the areas of the basins of attraction shown in Figure 3.1 with $\alpha = 0.7$, $\epsilon = 0.5$ and using a 400×400 grid. The numbers 8, 16, 32 and 64 refer to the number of iterates tested. Note that there is no large change in area as ν crosses 1 (loss of asymptotic stability), but near $\lambda = 1.3$ (extrapolating the number of iterates tested to ∞) the area of the basin goes to zero. At this point, there is a subcritical blowout bifurcation of the attractor.

Preimages of the fixed point 0 are dense in A ([68]) and so

$$T = \bigcup_{n \geq 0} f^{-n}(T_0)$$

is an open set in $\mathbb{R}^2$ whose boundary is dense in A and which consists of points which are repelled away to ∞. In other words, A is densely filled of points which are the bases of repulsive tongues. Therefore the basin $\mathcal{B}(A)$ is riddled with that of the attractor at infinity. It can be shown that, if we take a small neighbourhood U of A, the relative (Lebesgue) measure of the tongues T (i.e. $\frac{\ell(T \cap U)}{\ell(U)}$, where ℓ is 2-dimensional Lebesgue measure) converges to 0 with $\ell(U)$. This implies A is an essential attractor.

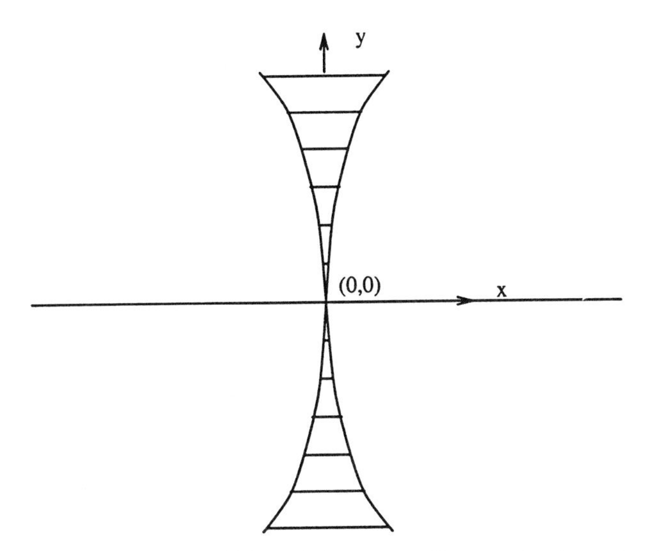

Figure 3.4: Local unstable manifolds of the origin. "Tongues" of instability consisting of points that are repelled away to infinity have this structure, and density of preimages of the origin in A implies that the preimages of the tongues form an open dense subset in any neighbourhood of the origin. Their measure, although positive, may be arbitrarily small.

(3) From (3.43) it follows that, for $\nu > e^{K_{SBR}\alpha}$, Λ_{SBR} is positive. It thus follows from Theorem 3.2.18 that $\mathcal{B}(A)$ has zero measure and therefore A is a chaotic saddle.

(4) By Theorem 3.2.16, if $\lambda_{\min} = \inf_{\mu \in Erg_f(A)} \lambda_\mu > 0$, there exists a neighbourhood V of A such that, if $|y| \neq 0$, then some iterate of (x, y) leaves V. The set A is then a normally repelling chaotic saddle. □

Remark 3.4.2 The essential point in part (2) is that the preimages of the fixed point $(0,0)$ are dense in A, and therefore the generic set G_μ for the ergodic measure $\mu = \delta_{(0,0)}$ is dense in A. We could then have appealed to Theorem 3.2.15 – but we find that the explicit construction of the repulsive tongues is helpful for the intuition. These tongues are the non-linear counterpart of the open sets constructed in Theorem 3.2.15. □

For both f and g, the loss of asymptotic stability, blowout bifurcation and the bifurcation to normal repulsion take place on the curves $\nu = 1$, $\nu = e^{K_{SBR}\alpha}$, and $\nu = e^{K_{\max}\alpha}$ respectively. This is shown in the (α, ν) plane in Figure 3.5 (p. 102); the lower and upper lines correspond to loss of asymptotic stability and bifurcation to normal repulsion respectively. The curve marked by the crosses corresponds to the blowout bifurcation.

We can picture the contents of the previous theorem in the following way. The bifurcation from (global) asymptotically stable attractor happens at $\nu = 1$

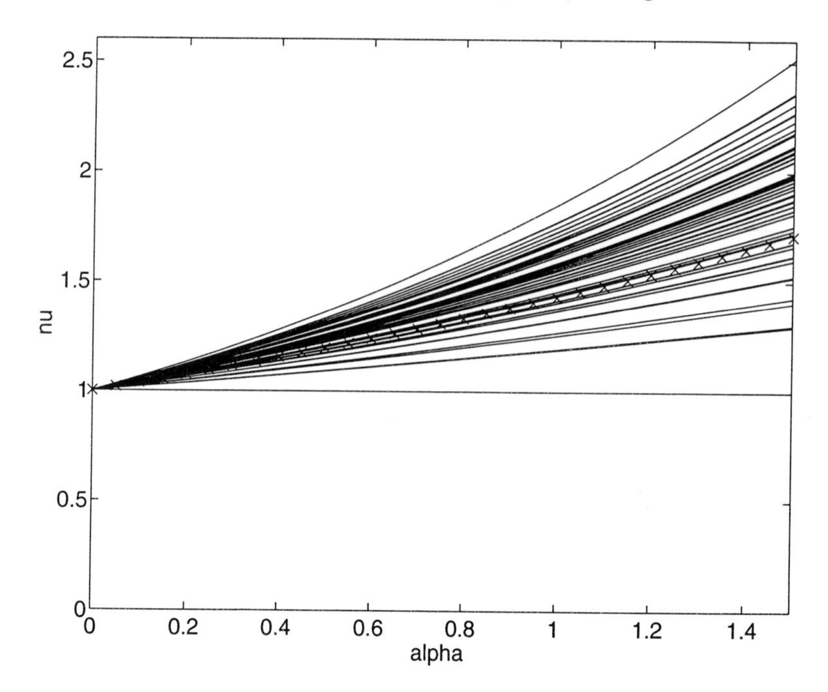

Figure 3.5: Bifurcations of f and g in the parameter plane (α, ν) (independent of ϵ). The line $\nu = 1$ corresponds to loss of asymptotic stability as the fixed point at the origin loses stability. The line indicated by crosses shows the blowout bifurcation of the natural measure. The uppermost line shows the bifurcation to normal repulsion corresponding to loss of normal stability of the symmetric period two orbit. Periodic orbits up to period seven are shown.

– the point at which $\Lambda_{\max}$ becomes positive. However, for $1 < \nu < e^{K_{SBR}\alpha}$ the natural measure μ_{SBR} remains normally stable (that is, $\Lambda_{SBR} < 0$); this implies that, in a measure-theoretic sense, "most" of the points in a small neighbourhood of A avoid the dense set of repulsive tongues and are attracted towards A.

As ν increases, progressively more ergodic invariant measures lose normal stability. As these can only be supported on periodic orbits or Cantor sets (by Proposition 1.3.3 and Theorem 1.3.14), the same phenomenon of creation of more repulsive tongues happens for these sets.

A significant change occurs when Λ_{SBR} becomes positive. Almost all points (in the Lebesgue measure sense) in A are now normally repulsive and we expect almost all points in a two-dimensional neighbourhood of A to be repelled away. However, there are still stable ergodic measures μ supported in A, with respect to which local 1-dimensional stable manifolds exist a.e.. Therefore, even in this situation A can attract a zero-measure set in a neighbourhood of itself.

Finally, after the last ergodic measure in A has become unstable at the point $\nu = e^{K_{\max}\alpha}$ there are no points in A with a transverse stable manifold. The only points which remain within a small neighbourhood of A are those in $y = 0$.

We note that periodic points in A become normally unstable through subcritical pitchfork bifurcations.

3.4.3 Global transverse stability for g

The case for g is somewhat more complicated to handle. Figure 3.6 (p. 104) shows a sequence of attractors for the map g, also at $\epsilon = 0.5$ and $\alpha = 0.7$.

As before, for $0 \leq \nu < 1$ A is an asymptotically stable attractor. For $\nu > 1$, g has a locally riddled basin attractor. Numerical experiments suggest that, at least for ϵ small enough, there are no attractors away from the invariant subspace; in this case $\mathcal{B}(A)$ is open but locally riddled. We abstract this in the following conjecture.

Conjecture 1 *For any $\alpha > 0$ and small enough $|\epsilon|$ the behaviour of the map g is as follows:*

1. *For $1 < \nu < e^{K_{SBR}\alpha}$, A is a Milnor attractor and $\mathcal{B}(A)$ is open.*
2. *For $e^{K_{SBR}\alpha} < \nu$ there is a family of attractors containing A with an SBR measure μ_ν that converges in the weak* topology to the SBR measure of $g_{|N}$ as ν tends to $e^{K_{SBR}\alpha}$ from above.*

3.5 Example: synchronization of coupled systems

An important problem for the dynamics of coupled systems is to obtain conditions ensuring synchronization. This has been studied by several authors; see for example Yamada *et al.* [120], Pikovsky *et al.* [91], Pecora *et al.* [87]. Suppose two identical systems are coupled in a way that preserves the symmetry obtained by interchanging the two systems. Then there will be an invariant subspace corresponding to the synchronized state, where the two systems trace exactly the same trajectory at the same time. This is sometimes called, for obvious reasons, the *synchronization manifold*. In this section we describe an experiment investigating the synchronization of two identical chaotic units as an application of the theory in the previous sections. Although we can only examine transients and stable dynamics, we find the characteristic signs of locally riddled basin attractors, as described in § 3.2: a normal spectrum extending across zero and a natural measure whose normal exponents are negative.

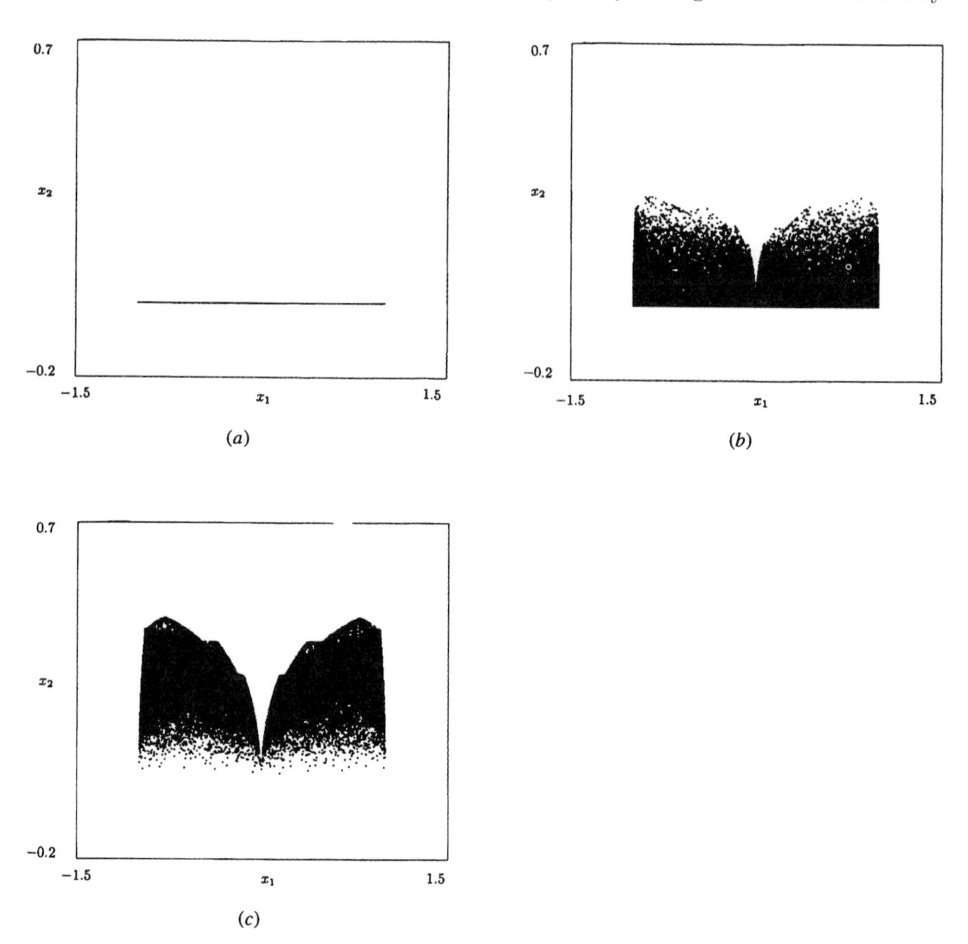

Figure 3.6: Attractors for $g_{\alpha,\nu,\epsilon}$ with $\alpha = 0.7$, $\epsilon = 0.5$ and increasing ν in the region $(x, y) \in [-1.5, 1.5] \times [-0.2, 0.7]$. Any transients were allowed to die away before these points were plotted.

(a) $\nu = 1.2$; an attractor with a locally riddled basin exists.

(b) $\nu = 1.3$; the attractor has undergone a blowout to create on-off intermittent behaviour of an attractor with no symmetry; the invariant measure is strongly concentrated near the x-axis (100,000 iterates).

(c) $\nu = 1.4$; the deviations of the trajectory away from the invariant subspace are more frequent (50,000 iterates).

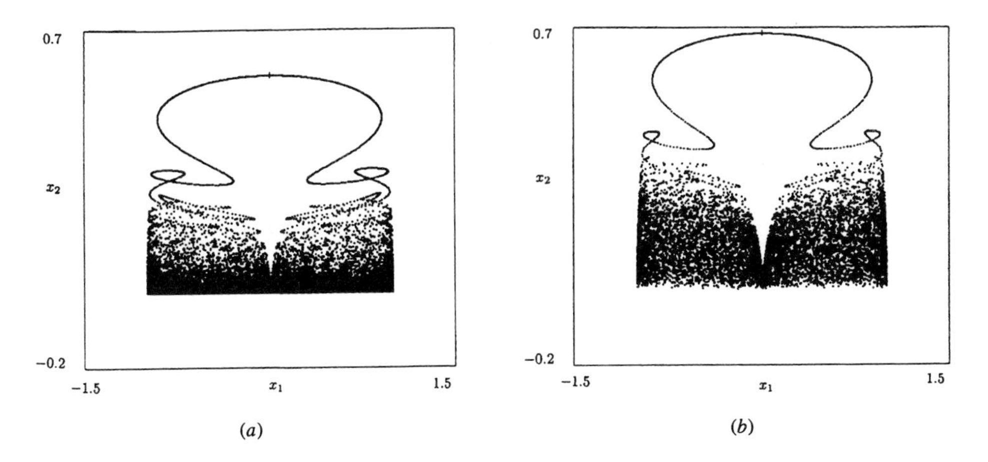

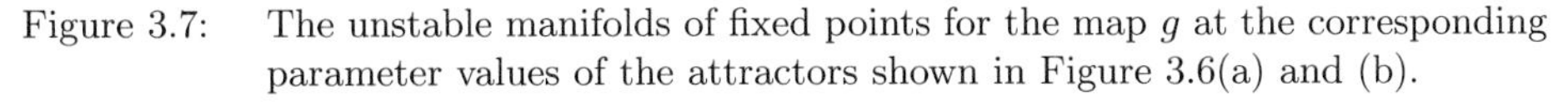

Figure 3.7: The unstable manifolds of fixed points for the map g at the corresponding parameter values of the attractors shown in Figure 3.6(a) and (b).

(a) At $\nu = 1.2$ the manifolds appear to be dense in a neighbourhood of the attractor with the locally riddled basin.

(b) At $\nu = 1.3$ the manifold has changed little, but the attractor has changed to apparently include this unstable manifold.

3.5.1 Electronic experiments

To examine the phenomena described in § 3.2 and § 3.3 we consider an electronic system of two coupled three degree of freedom chaotic oscillators. The circuit of one chaotic oscillator is shown in Figure 3.8 (p. 106). This consists of five 7611 operational amplifiers configured as two inverters and three integrators. Nonlinearity is provided by two OA202 diodes which prevent saturation of the operational amplifiers. The supply is regulated at ± 5V. The circuit has an attracting limit cycle for small values of R which undergoes a period-doubling cascade to chaos, producing a Rössler-type attractor for larger values of R. We shall use the fact that this attractor is well approximated by a unimodal map on an interval. We assume a non-linearity $I = f(V)$ models the diode response.

The components used are of low tolerance and high temperature stability to ensure the behaviour of the systems is near-identical. In particular, 1% tolerance resistors and 2.5% tolerance capacitors are used; the $20nF$ capacitors are constructed from two matched $10nF$ capacitors in parallel. Ten-turn trimmers are used to adjust R_a in one oscillator and R_b in the other. The two oscillators are set up by adjusting these trimmers when both oscillators are in the same state (on the point of undergoing a 3-to-6 period doubling) to ensure both frequency and amplitude are identical. Care is taken to ensure the room temperature is kept at 20^oC to within one degree and the construction is such that air can freely circulate within the enclosure.

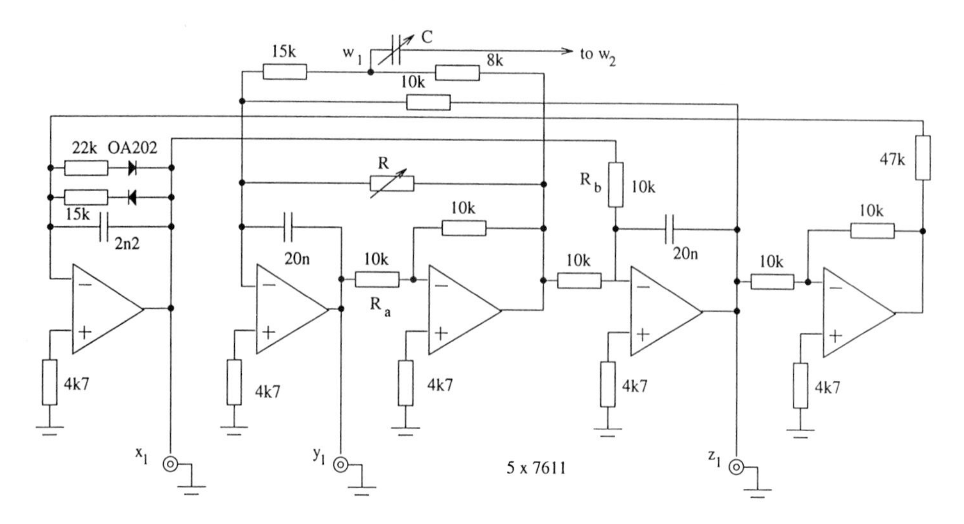

Figure 3.8: Circuit diagram of a three degree of freedom chaotic oscillator. By varying R it is possible to observe a breakdown to chaos via period doubling of a limit cycle, giving a Rössler-type attractor. Coupling is via the capacitor C, which introduces an extra degree of freedom and is a normal parameter for the synchronized state.

Coupling is provided by breaking a feedback loop and connecting via a capacitance decade box. This means that the coupled system has seven degrees of freedom. The results for the region of interest show that the asymptotic flow is effectively on a four dimensional branched manifold – product of two Rössler-type attractors. The full equations are given by

$$\left\{\begin{array}{rcl} \dot{x}_1 & = & \dfrac{z_1}{103.4} + f(x_1) \\ \dot{y}_1 & = & \dfrac{y_1}{22R} - \dfrac{z_1}{200} - \dfrac{w_1}{300} \\ \dot{z}_1 & = & -\dfrac{x_1}{200} + \dfrac{y_1}{200} \\ C(\dot{w}_1 - \dot{w}_2) & = & -\dfrac{w_1}{15} - \dfrac{y_1 + w_1}{8} \end{array}\right.$$

with a similar equation for the second chaotic oscillator, obtained by interchanging the subscripts 1 and 2. C is measured in nF and time in milliseconds. By setting $w = w_1 - w_2$ it is possible to write this as an ordinary differential equation on $\mathbb{R}^7$ by setting

$$w_1 = \frac{w}{2} - \frac{8}{46}(y_1 + y_2).$$

As $C \to 0$, the equation for $\dot{w}$ becomes singular with unique solution

$$w_1 = -\frac{15}{23} y_1,$$

reducing to two third order uncoupled equations. We also note that C is a normal parameter as in Definition 3.3.2, since the dynamics in the synchronized state are independent of C.

The action of $\mathbb{Z}_2$ by permuting the oscillators gives an action on the phase space $\mathbb{R}^7$ by permuting the subscripts and reversing the sign of w. The fixed point space for this action, or *space of synchronized states,* is given by $x_1 = x_2$, $y_1 = y_2$, $z_1 = z_2$ and $w = 0$.

3.5.2 Observations

Observations of the dynamics of the circuit are made by plotting x_1 against x_2 on an oscilloscope. Both circuits are set running in the same chaotic state with $R = 39.5k\Omega$ and the coupling capacitance is varied from $C = 0pF$ up to $C = 10nF$. For low values of the coupling, the lack of synchronization of the oscillators shows in the $x_1 - x_2$ plane as a seemingly random motion over the whole of a square; see Figure 3.9(a) (p. 108). From $C = 30pF$ there is evidence of some degree of synchronization; this is shown for $C = 210pF$ in Figure 3.9(b). This gives way via a reverse cascade of period doublings to a stable periodic orbit at $C = 1100pF$ that is not in-phase and is pictured in Figure 3.9(c). This persists up until C is $1139pF$, where it loses stability to a synchronized chaotic attractor confined to the line $x_1 = x_2$. Decreasing the coupling capacitance by a small amount shows that this synchronized attractor persists hysteretically with the non-synchronized periodic orbit. For C lower than about $1400pF$, small deviations away from the line $x_1 = x_2$ are noticeable at irregular intervals. For C reduced further to about $1090pF$, the in-phase chaotic attractor is only observable as a transient, although it can persist for up to a minute. It is on this state of weak stability that we concentrate our attention.

To examine the dynamics more closely, we take samples simultaneously from x_1, y_1, x_2, y_2 at a rate of $12.5kHz$ using a 16-bit analog to digital converter. This records time series from the four channels simultaneously and allows easy reconstruction of the dynamics.

3.5.3 Analysis of the dynamics

For the case of a synchronized attractor exhibiting occasional small excursions from the synchronized subspace, we now reconstruct the dynamics and some of its invariant measures restricted to the subspace $x_1 = x_2$, $y_1 = y_2$. Figure 3.9(d) shows a segment of the time series (x_1, y_1, x_2, y_2) for the parameter values $R = 39.5k$ and $C = 1110pF$. The attractor is close to a Rössler attractor and the line $x_1 + x_2 = 1.0V$ is a reasonable codimension one section for the dynamics. The same section is a codimension one *equivariant* section for (x_1, y_1, x_2, y_2) measured as time series, and induces a return map on the three dimensional space $(p, d_x, d_y) = (y_1 + y_2, x_1 - x_2, y_1 - y_2)$. We denote

$$d = \sqrt{d_x^2 + d_y^2}$$

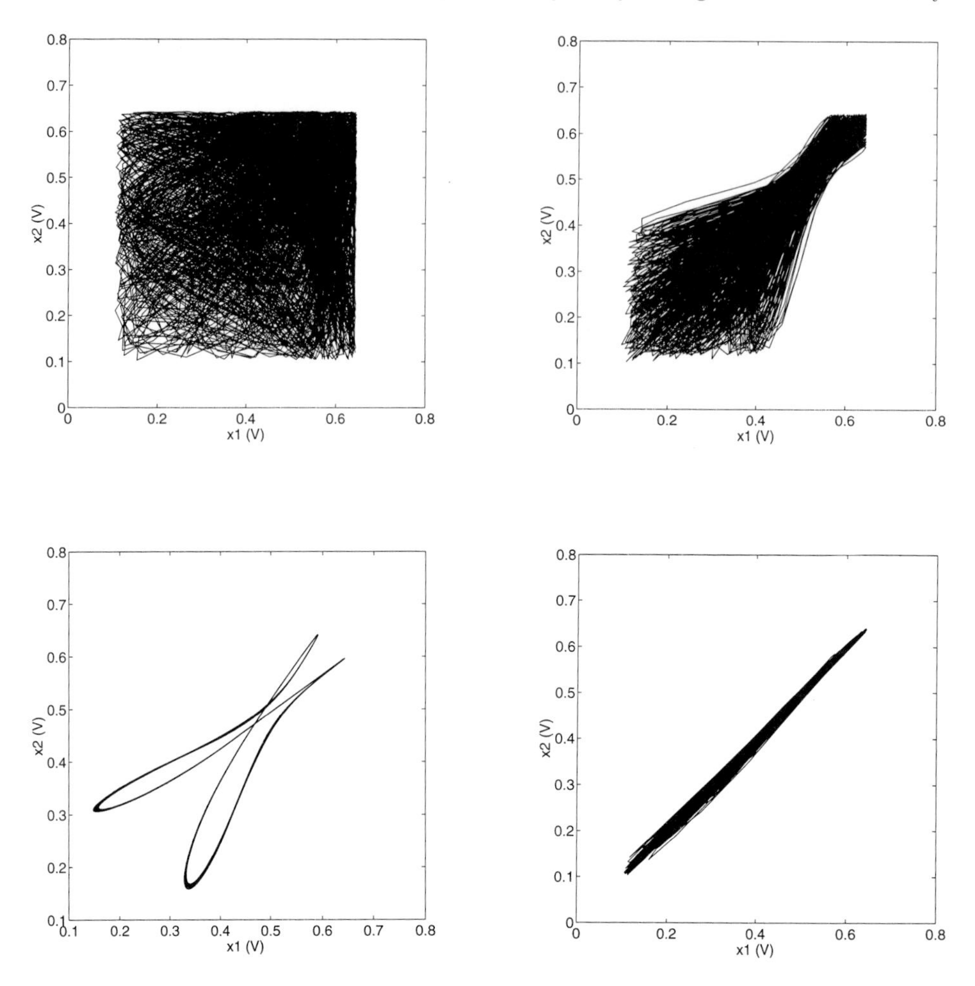

Figure 3.9: Plotting the voltages x_1 against x_2 for a range of coupling capacitances. All figures were produced with $R = 39.5k$.

(a) (upper left) $C = 0$, uncoupled. There appears to be no long-time correlation between the units.

(b) (upper right) $C = 210pF$: the units are synchronized on average but, at any moment in time, they behave differently.

(c) (lower left) $C = 1100pF$: the units have a stable periodic orbit that is not synchronized.

(d) (lower right) $C = 1110pF$: the system evolves towards a state which is close to synchronization for most of the time, but makes sudden deviations away from it in a seemingly irregular way (bubbling). Plotted in the (x_1, x_2) plane, the transients going away from the synchronized state are visible as deviations of the trajectory from the line $x_1 = x_2$.

the deviation from the fixed point space and

$$\theta = \tan^{-1} \frac{d_y}{d_x}.$$

The section is taken by observing crossings by the continuous time series and interpolating all observables using a spline smoothing to give variables at one tenth of a sample length, then linearly interpolating between the smoothed trajectory to obtain returns on the section. The fixed point space of the $\mathbb{Z}_2$ action in this space is given by $d_x = d_y = 0$, that is, the p-axis. The $\mathbb{Z}_2$ action acts as a rotation by π about this fixed point space.

We wish to approximate the ergodic invariant measures on the fixed point space and find the associated normal Liapunov exponents. This is done in four stages; firstly, we reconstruct the dynamics in the fixed point space. Secondly, we reconstruct linearized dynamics for the direction normal to the fixed point space. Thirdly, we approximate periodic orbits and the natural measure for the dynamics on the fixed point space. Finally, we approximate the normal Liapunov exponents by integrating the normal dynamics with respect to the different invariant measures obtained.

Reconstructing the fixed point space dynamics

Since this is an experiment, we cannot ensure that the two oscillators are exactly identical; we can only assume the difference between them is small. We introduce two small numbers, $0 < d_{\min} < d_{\max}$ such that

(i) The fixed point space is contained within $d < d_{\min}$.
(ii) The behaviour within $d < d_{\min}$ in the p direction is determined by that on the fixed point space.
(iii) For $d_{\min} < d < d_{\max}$ the behaviour normal to the fixed point space is dominated by a linear map.
(iv) A significant number of points lie within the range $(d_{\min}, d_{\max})$.

For the sample in Figure 3.9(d) we take $d_{\min} = 0.01$ and $d_{\max} = 0.02$. For the measured return map $(p(n), d_x(n), d_y(n))$ for $n = 1, \ldots, N$ we plot $p(n+1)$ against $p(n)$ for those n such that $d < d_{\min}$. The return map shown in Figure 3.10 (p. 110) is obtained for the sample discussed. The plot includes a fitted 6th order polynomial return map. To a very good approximation $p(n+1)$ is determined by $p(n)$, apart from a small amount of random noise with standard deviation $0.47mV$; this supports (i) and (ii) of the above requirements for d.

Reconstructing the linearized local dynamics

We assume that a linear map dominates the behaviour of returns in the annulus $d_{\min} < d < d_{\max}$. For points landing in this annulus, we plot the angle θ from the plane $d_y = 0$ in the (d_x, d_y) plane. As seen in Figure 3.11 (p. 111), this

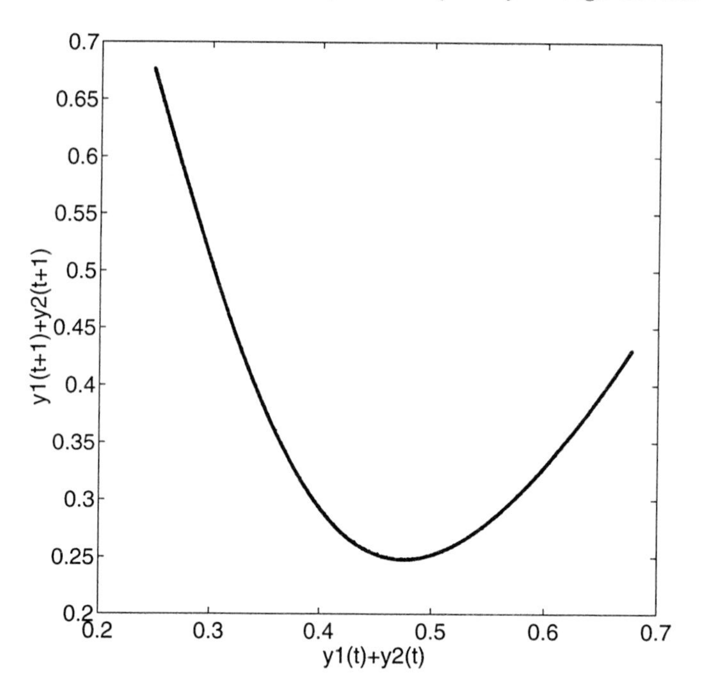

Figure 3.10: Return map obtained by interpolating the trajectory shown in Figure 3.9(d) and restricting to a tubular neighbourhood of the synchronization with radius $5mV$. A sixth order polynomial curve is fitted to this return map; the rms error is $0.47mV$.

is concentrated in one band (modulo π), implying that the dynamics split into a strongly attracting direction and a marginally stable direction. In particular, this suggests that the linearization near the invariant subspace satisfies a cone condition ensuring continuity of the Liapunov exponents. We concentrate only on those points in the annulus with $\theta_{\min} < \theta < \theta_{\max}$ for $\theta_{\min}$ and $\theta_{\max}$ chosen to exclude any outliers. We now make the assumption that the dynamics in this marginally stable direction are determined by the dynamics in the d-direction only, and the dynamics here is close to a linear map.

Thus plotting $\log d(n+1) - \log d(n)$ for n such that $d_{\min} < d(n) < d_{\max}$ and $\theta_{\min} < \theta(n) < \theta_{\max}$ we obtain Figure 3.12 (p. 112). We can now support our assumptions (iii) and (iv) for the choice of $d_{\min}$ and $d_{\max}$ by assuming a linear map

$$L(d(n)) = d(n+1)$$

and observing that these points lie close to a line which we approximate using a 6th order polynomial least-squares fit. Note that $\log L > 0$ corresponds to local expansion, $\log L < 0$ to local contraction.

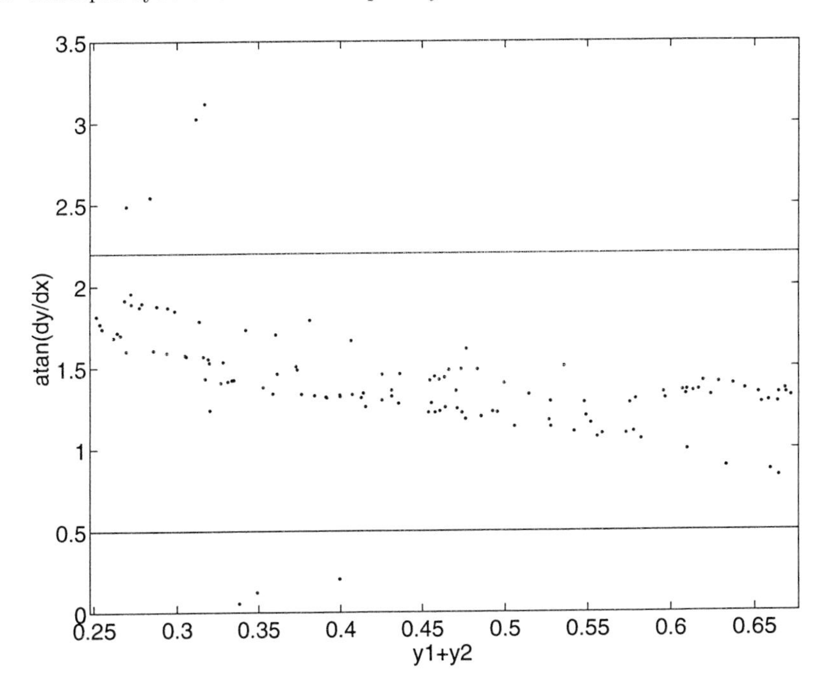

Figure 3.11: Near the synchronized state, this plot shows the angle at which the successive returns arrive in the (d_x, d_y)-plane. Note that they are concentrated in a band suggesting that one continuously defined direction corresponds to the least stable direction of the linearization of the dynamics on the synchronized state.

Invariant measures for the reconstructed map

There are two easily approximated classes of invariant measures for the interval map on the fixed point space. These are the natural measure and the periodic point measures. The former can be obtained from box counting (using 50 boxes) the points on the time series which lie within $d < d_{\min}$ in the p direction to give the measure $\mu_1(y)$; see Figure 3.13 (p. 113). However, care needs to be taken as this only counts parts of the orbit which are near the fixed point space and could be affected by the dynamics normal to the fixed point space. In other words, what we get by box-counting the points with $d < d_{\min}$ is an approximation to the conditional measure on the fixed point space of some natural measure and not the invariant measure of the map restricted to the fixed point space. For comparison we examine the invariant measure obtained from the *reconstructed map* on the fixed point space. Suppose σ is the standard deviation of the samples in Figure 3.10 from the fitted curve $f(p)$. We approximate the natural measure on the fixed

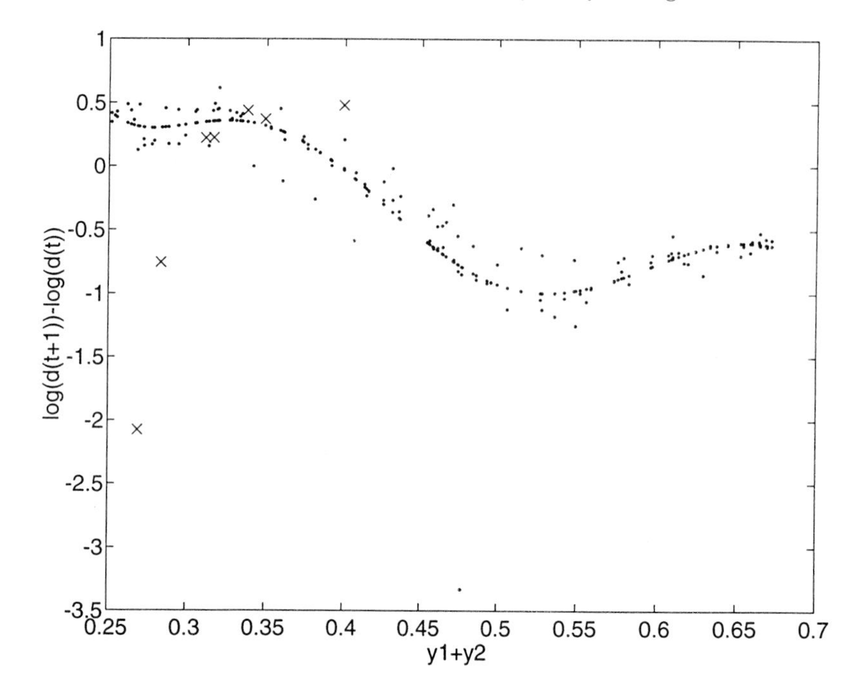

Figure 3.12: On the least stable manifold, the points show the logarithm of the ratio of successive returns, $\log L = \log d(t+1) - \log d(t)$, where t is such that $0.5 < \theta(t) < 2.2$ and $0.005 < d(t) < 0.01$. A positive exponent indicates local repulsion, whereas a negative exponent indicates local attraction. Data are fitted to a sixth order polynomial using a least squares fit. The crosses correspond to samples where $\theta(t)$ is not in the range $0.5 < \theta(t) < 2.2$.

point space using the random map

$$p_{n+1} = f(p_n) + \sigma r_n,$$

where r_n is a uniformly distributed random variable in $[-1/\sqrt{3}, 1/\sqrt{3}]$ (r_n has zero mean and standard deviation 1/3). Iterating this random map and box counting using 50 boxes gives the normalized invariant measure $\mu_2(y)$ also shown in Figure 3.13.

The periodic invariant measures were obtained by searching for periodic points of f of period smaller or equal to 6). This was done using a binary chop interval search to locate roots of $f^p(y) - y = 0$.

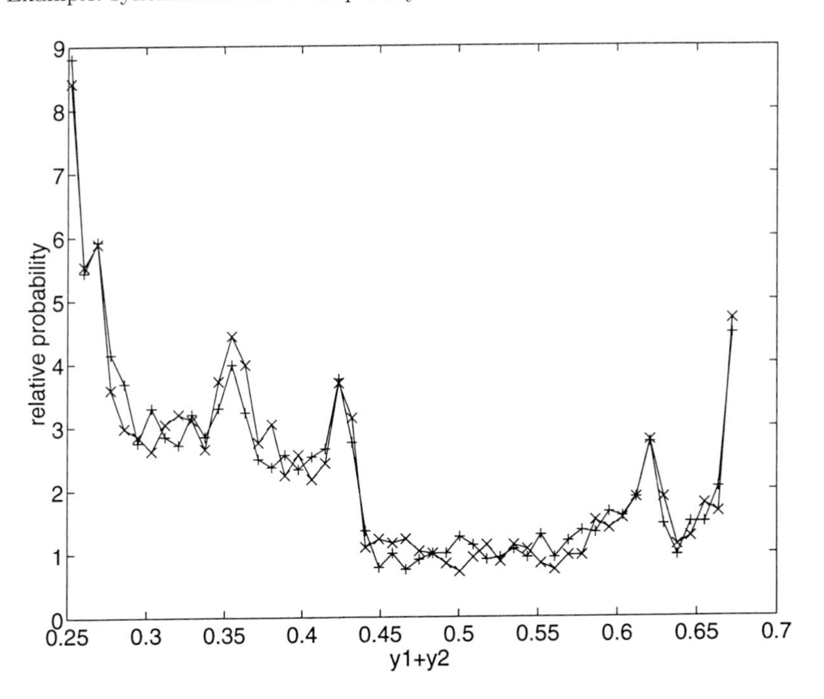

Figure 3.13: The measured and observed invariant probability densities for the map on the synchronized space shown in Figure 3.10. The measure μ_1 marked by 'x' refers to the invariant probability measure calculated by box-counting 3600 points from iterating the fitted return map and adding a small random perturbation. The measure μ_2 marked by '+' is obtained by box-counting the 3600 returns satisfying $d < d_{\min}$.

Normal Liapunov exponents

The largest normal Liapunov exponent for the natural measure approximated by μ_1 or μ_2 is

$$\Lambda_{SBR} = \int_y \log(L(y))d\mu(y), \tag{3.45}$$

while for a periodic orbit $\{y_1, \dots, y_k\}$ of period k it is given by

$$\Lambda_p = \frac{1}{k}\sum_{i=1}^{k} \log(L(p_i)). \tag{3.46}$$

Using (3.45) and (3.46), Figure 3.14 (p. 114) shows, for several invariant measures μ, a plot of Λ_μ against supp μ. Note that the largest Liapunov exponent $\Lambda_{\max} = 0.277$ is positive and corresponds to Dirac measure on the fixed point of the map f. The computed Liapunov exponents for both approximations of the

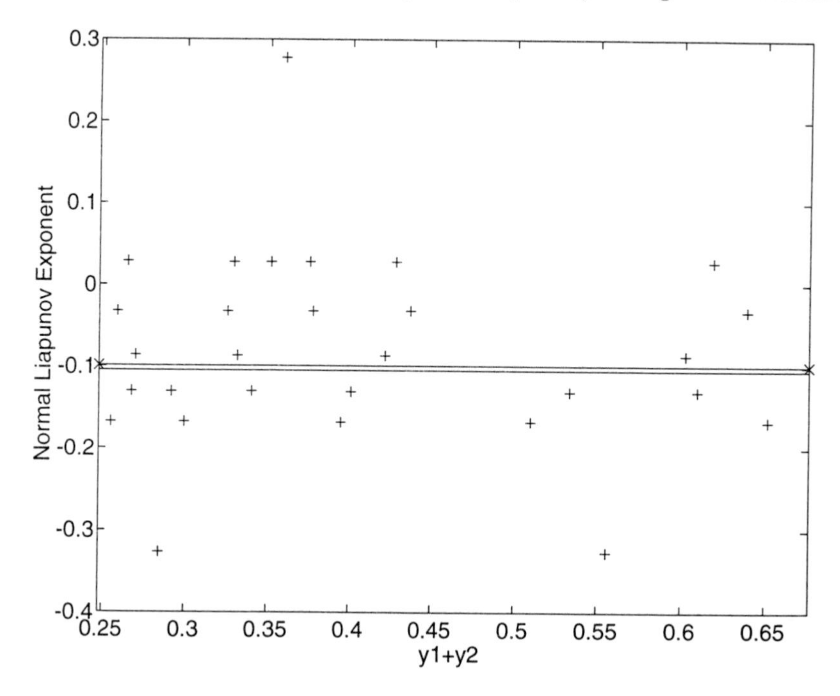

Figure 3.14: The normal Liapunov exponents calculated for the two approximations of the natural invariant measure, and Dirac measures supported on periodic orbits. The horizontal line ending in 'x' is the normal Liapunov exponent for μ_2, while that without 'x' corresponds to μ_1. Note that $\Lambda_{\max} > 0 > \Lambda_{SBR} > \lambda_{\min}$, indicating that the deviations away from the synchronized state are 'bubbling' caused by low level noise in the system.

natural measure, $\Lambda^{(1)}_{SBR} = -0.2442$ for μ_1 and $\Lambda^{(2)}_{SBR} = -0.2304$ for μ_2, are negative – as expected by the observed attractivity of the fixed point space. Moreover, there are periodic points (of period less than 6) whose normal Liapunov exponents are smaller than Λ_{SBR}. The smallest such exponent is $\Lambda_{\min} = -0.3264$ and corresponds to a period 2 orbit.

We can estimate the errors of the measured Λ_{SBR} from the true Λ^*_{SBR} by using the fact that box counting gives a standard deviation of $1/\sqrt{N}$, where N is the average number of samples which lie in a box. Using the 50 boxes and approximately $4,000$ samples to obtain the reconstructed natural measure we obtain $N = 80$, giving a standard deviation of about 0.11 on the measure assigned to each box. The error of the measured Liapunov exponents can be estimated if μ and its approximation μ^* are absolutely continuous. From Figure 3.12 we consider $\mu(p) = \mu^*(p) + \tilde{\mu}(p)$ with $\tilde{\mu} \in L^1$ representing the error. Then

$$|\Lambda^*_{SBR} - \Lambda_{SBR}| \leq \left| \int \log L \tilde{\mu} dp \right| \leq \left| \int \left[\frac{d}{dp}(\log L) \int^p \tilde{\mu}(s) ds \right] dp \right| .$$

We can estimate the derivative of $\log L$ to be bounded by 4, and noting that $\int^p \tilde{\mu}$ is bounded by $1/\sqrt{N}$ we get

$$|\Lambda^*_{SBR} - \Lambda_{SBR}| \leq \left\| \frac{d}{dp} \log L \right\| \int^p \tilde{\mu}\, dp < \frac{4}{\sqrt{4000}} \approx 0.06.$$

This is comfortably smaller than the measured Λ_{SBR}. A larger source of uncertainty comes from the fit of L in Figure 3.12. There are several sources of the scatter of data (rms deviation $= 0.27$) from the fitted curve. Firstly, noise will be more noticeable on the direction of weaker Liapunov exponents. Secondly, the measurement of d from the fixed point space is adversely affected by the fact that the systems are not identical. Thirdly, there may be points we are fitting to the curve which are not in the 'least stable manifold'. Finally, the linear approximation may be not a very good one in the selected region of d. We believe that the first and second sources of error will be the main ones.

3.6 Historical remarks and further comments

Remark 3.6.1 The subject dealt with in this chapter is a very rich one and has recently attracted much attention from several directions. In what follows we try to sketch the recent evolution of concepts and ideas.

One of the areas in which phenomena related to the blowout bifurcation were first observed is that of Symmetric Chaos, particularly in relation to symmetry increasing bifurcations. In 1988 Chossat *et al.* notice the occurrence of a blowout bifurcation – even though, to be exact, the concept did not even exist – in numerical experiments with symmetry-increasing bifurcations of a D_3-symmetric map. In one of the observed scenarios, a chaotic one-dimensional attractor in the fixed-point space $\mathrm{Fix}(\mathbb{Z}_2)$ 'explodes' into a much larger $\mathbb{Z}_2$-symmetric set. King and Stewart [56] observe a similar phenomenon in connection with "symmetry-increasing crises".

Rand *et al.* [96] examine evolutionary dynamics, where the dynamics is a map on a simplex that maps the boundary components of the simplex (representing zero population levels) to themselves. They develop a theory of the invasion of the ecology by a mutant species through what they call the *invasion exponents* – essentially our Λ_{SBR}. With some simplifying assumptions, they state a classification theorem related to our Theorem 3.2.12.

A third direction in which such problems arose naturally was that of synchronization of chaotic systems. The work of Pecora and Carroll [87] in 1990 was, in a sense, groundbreaking. They split a chaotic system into two subsystems, A and B; two subsystems B driven by the same system A will typically synchronise only if the Liapunov exponents of the B subsystems are negative. In 1991, Pikovsky and Grassberger [91] investigate a system of two coupled tent maps and note that even when a system has a stable synchronised state indicated by the largest Liapunov exponent, the basin of attraction may be densely filled with periodic points; the attractor is surrounded by a strange invariant set which is dense in a neighbourhood of the attractor. This attractor can therefore only be a Milnor attractor. They also observe a bifurcation of this attractor to one containing this larger invariant set.

In 1992 Alexander *et al.* [1] for the first time observed and defined the phenomenon of riddled basins. They observed the connection with normal exponents and proved Theorem 3.2.17. Subsequently, Ott *et al.* [83] found that the same behaviour near the invariant subspace can give rise to two different scenarios; either intertwined riddled basins, or on-off intermittency, originally defined by Platt *et al.* [93], where excursions away from the attractor are globally re-injected. Ott *et al.* defined the point at which the largest transverse Liapunov exponent goes through zero to be a *blowout bifurcation.*

These works sparked a wave of interest; the area itself has since undergone a blowout. Many exploratory papers in several aspects of blowout bifurcations and on-off intermittency have since appeared, most of them in the experimental or applied side. Numerical and experimental systems exhibiting locally riddled basins and/or on-off intermittency are studied in Ashwin *et al.* ([4], [5]), Hammer *et al.* [43]. The effect of noise in on-off intermittency is considered in Ashwin *et al.* [4], Platt *et al.* [92], and studies on aspects of scaling are reported in Heagy *et al.* [44]. Scaling properties of such systems are also treated in Ott *et al.* [81]. Behaviour with parameters when approaching a blowout bifurcation is treated by Ott *et al.* [81]. Kan [53] exhibits an *open* set of diffeomorphisms of the thickened two-torus with boundary with two different attractors whose basins are intermingled – in fact everywhere dense on the manifold. Heagy *et al.* [45] perform an experiment with a system of synchronized chaotic circuits, similar in spirit to the one described in § 3.5.1. This system was *a priori* known to have several coexisting attractors away from the synchronization manifold, thus allowing for direct experimental observation of riddled basins. Lai *et al.* [59] investigate the characterization of riddled fractal sets, with a particular emphasis on the distribution of normal exponents on the attractor A. In the case of a riddled basin, these fluctuate widely in sign. They characterize these fluctuations by the introduction of sign-singular measures and the associated cancellation exponent. Venkataramani *at al.* [117] investigate the effects of noise and asymmetry on the bubbling transition, concluding that the attractor will generically disappear (which is to be expected in view of Remark 3.2.21) and that near its "ghost" there is either a chaotic transient or an intermittently bursting time evolution, deriving scaling relations in both cases.

Remark 3.6.2 (Criticality of the blowout bifurcation) The evidence for the maps f and g in § 3.4 suggests that it is possible to define 'criticality' for blowout bifurcations, analogous to that for bifurcation of fixed points in invariant subspaces. For example, steady state bifurcations with $\mathbb{Z}_2$ symmetry are generically pitchfork bifurcations that are either subcritical or supercritical. This criticality corresponds to the hysteretic and non-hysteretic scenarios observed by Ott *et al.* [83].

For f, we note that there is an unstable invariant set A_ν, namely the boundary dividing the basins of A and the attractor at infinity for $\nu < \nu_0$, and this is destroyed on passing through $\nu = \nu_0$. In this sense, f exhibits a *subcritical* blowout bifurcation on passing through ν_0.

For g, there is numerical evidence of a family of attractors $\{A_\nu : \nu > \nu_0\}$; these correspond to on-off intermittent attractors. We conjecture that (at least for ϵ and $\nu - \nu_0$ small enough) there exists no other attractor whose basin could riddle that of A. Presumably the attractors in A_ν are chaotic only for a set of parameters $\nu > \nu_0$ of positive measure. Although A_ν appears to approach $\tilde{A}$ (the closure of the union of unstable manifolds of A) as ν approaches ν_0, the natural invariant measures μ_ν supported on A_ν appear to satisfy

$$\mu_\nu \to \mu_{SBR} \quad \text{as} \quad \nu \to \nu_0 \text{ from above,}$$

with convergence in the weak* topology. Thus we say that g exhibits a *supercritical* blowout bifurcation. Figure 3.7 show examples of the unstable manifolds from the fixed points that have bifurcated from the origin.

Thus, there appears to be an 'important' family of invariant sets that appear to branch from A for only $\nu < \nu_0$ in the subcritical case or $\nu > \nu_0$ in the supercritical case. For the supercritical case, we observe that these invariant sets are attractors.

Note that for g, all periodic points in A are observed to undergo supercritical pitchfork bifurcations on varying ν, whereas those for f are all subcritical; thus we might expect that all bifurcations of A for g are supercritical while for f they are subcritical. However, this need not generally be the case. We emphasise that as yet we have no proof that the criticality for the blowout bifurcation is well-defined.

In contrast to the situation for steady-state bifurcation where the bifurcation is determined by dynamics on any small neighbourhood around the fixed point, for blowout bifurcations we need to consider dynamics in a neighbourhood of $\tilde{A}$ to discover the criticality of the bifurcation. Thus, 'higher order normal Liapunov exponents' (relative to higher order normal derivatives) for the attractor A will not determine criticality; dynamics far from A are immediately important.

Remark 3.6.3 (Blowout and the Kaplan-Yorke formula) We briefly mentioned that near the blowout bifurcation there will be no normal hyperbolicity and therefore the attractors will not persist on some perturbed manifold of the same dimension as the invariant manifold under a general, arbitrarily small perturbation of the map. Numerical experiments indicate that, on introducing perturbations that destroy the invariant submanifold, an attractor of higher dimensionality than that of the original invariant manifold will be created. This is suggested by the Kaplan-Yorke conjecture [54] which states that the Hausdorff dimension of an attractor is, under suitable assumptions, equal to the Liapunov dimension. Since a Liapunov exponent crosses zero at a blowout bifurcation, the Liapunov dimension can exceed the dimension of the original submanifold. On perturbing to break invariance of the submanifold, we expect the Kaplan-Yorke formula to hold; this will give a discontinuous change in the dimension of the attractor.

Remark 3.6.4 (Noise) As with perturbations that destroy invariance of the submanifold, addition of noise that does not preserve this manifold should give rise to a discontinuity in the size of the attractor near blowout bifurcation, see Ashwin *et al.* [4]. In particular, addition of noise to a map $f : M \to M$ with attractor A in an invariant submanifold N, $f(N) \subset N$ will have the following effects, see also [43], [81], [82], [83], [92]:

(i) If A is a Milnor attractor whose basin is riddled with that of an asymptotically stable attractor C, then addition of low level noise will cause all trajectories in a neighbourhood of A to converge to C almost surely.

(ii) If A is a Milnor attractor whose basin is locally riddled but open, then addition of low level noise will cause all trajectories in a neighbourhood of A to recurrently explore a neighbourhood of $\tilde{A}$, the union of all unstable manifolds of points in A. We call this behaviour *bubbling* of the attractor, and as such it resembles on-off intermittency. This sort of behaviour has been observed by Platt *et al.* [92] to cause a discontinuous change in the parameter where the blowout bifurcation appears.

In both cases, A is no longer an attractor after addition of noise. In the first case it disappears; in the second we expect to see intermittent excursions transversely away from A in a manner similar to on-off intermittency [93], with the difference that this is now a noise-driven phenomenon. Note that there have been experimental observations of on-off intermittency and observation of a scaling law for the distribution of laminar phases [43]. We expect there will be scaling properties of the invariant measure for the noisy problem that distinguishes it from noisy forcing of an asymptotically stable attractor (see [92]).

Remark 3.6.5 (Symmetry of attractors) Following Melbourne *et al.* [25], we define two subgroups of Γ that characterize the symmetry of an attractor A. They are the *symmetry on average*,

$$\Sigma_A = \{\sigma \in \Gamma \mid \sigma A = A\}$$

and the *pointwise symmetry*,

$$T_A = \{\sigma \in \Gamma \mid \sigma x = x \ \forall x \in A\}.$$

Note that T_A is a normal subgroup of Σ_A. For finite groups Γ, Melbourne *et al.* [25] and Ashwin and Melbourne [6] have classified the possible subgroups of Γ that can be realized by Σ_A assuming that $T_A = \mathbf{1}$. This has been extended by Melbourne [67] to the case where T_A is non-trivial (note that T_A must be an isotropy subgroup for the action of Γ).

In § 3.4, the attractors created at the blowout bifurcation for the map g appear to change symmetry in the following way:

1. For $0 \le \nu < e^{K_{SBR}\alpha}$ there is a single attractor A with $T_A = \Sigma_A = \mathbb{Z}_2$.
2. For $\nu > e^{K_{SBR}\alpha}$ there are two conjugate attractors A_i with $T_{A_i} = \Sigma_{A_i} = \mathbf{1}$.

Remark 3.6.6 (Concluding remarks) The work reported in this chapter was realized in most ways independently of the developments described in Remark 3.6.1. This is clear both in the methodology and in the results. We consider that the spectrum of normal Liapunov exponents lies closer to the heart of the matter than other approaches, some of which are *ad hoc*: by considering the Liapunov exponents of all ergodic measures supported on the attractor we explain the precise origin of the regions of normal instability near attractors and the nature of locally riddled basins.

This work suggests many directions for generalizations. The case of invariant manifolds of codimension higher than one should pose many interesting questions. Finding ways to characterise global stability would also be of interest; there could be crises where the unstable manifold from a point on A hits another invariant set. It would also be interesting to find out whether a formal connection exists, and if so in which conditions, between the blowout bifurcation and the Kaplan-Yorke formula.

Bibliography

[1] J.C. Alexander, I. Kan, J.A. Yorke and Zhiping You, Riddled Basins. *Int. Journal of Bifurcations and Chaos*, **2** (1992): 795–813.

[2] Ll. Alsedà, S. Ben Miled, J. M. Gambaudo and P. Mumbrú, Asymptotic linking of periodic orbits for diffeomorphisms of the 2-disk. *Nonlinearity* **6** (1993): 653–663.

[3] D. K. Arrowsmith and C. M. Place, *An Introduction to Dynamical Systems*, Cambridge University Press; Cambridge 1990.

[4] P. Ashwin, J. Buescu and I.N. Stewart, Bubbling of attractors and synchronisation of oscillators. *Physics Letters A* **193** (1994): 126–139.

[5] P. Ashwin, J. Buescu and I.N. Stewart, From attractor to chaotic saddle: a tale of transverse instability. *Nonlinearity* **9** (1996): 703–737.

[6] P. Ashwin and I. Melbourne, Symmetry groups of attractors. *Arch. Rat. Mech. Anal.* **126** (1994): 59–78.

[7] M. M. Barge and R. B. Walker, Nonwandering structures at the period-doubling limit in dimensions 2 and 3. *Trans. Am. Math. Soc.* **337** (1993), 1: 259–277.

[8] H. Bell, A fixed point theorem for plane homeomorphisms. *Bull. Am. Math. Soc.* **82(5)** (1976), 778–780.

[9] H. Bell, On fixed point properties for plane continua. *Trans. Am. Math. Soc.* **128** (1977), 778–780.

[10] H. Bell and K. Meyer, *Limit periodic functions, adding machines and solenoids. J. Dyn. Diff. Eqns.* **7** (1995), 3: 409–422.

[11] M. Benedicks and L. Carleson, On iterations of $1 - ax^2$ on $[-1, 1]$. *Ann. Math.* (2) **122** (1985), 1: 1–25.

[12] M. Benedicks and L. Carleson, The dynamics of the Hénon map. *Ann. Math.* (2) **133** (1985), 1: 73–169.

[13] M. Benedicks and L. S. Young, SBR measures for certain Hénon maps. *Inv. Math.* **112** (1993), 3: 541–76.

[14] L.S. Block and W.A. Coppel, *Dynamics in One Dimension.* Lecture Notes in Math. vol. 1513, Springer-Verlag, Berlin, 1992.

[15] L. Block and E. M. Coven, ω-limit sets for maps of the interval, *Ergod. Th. Dynam. Sys.* **6** (1986): 335–344.

[16] R. Bowen, A horseshoe with positive measure. *Invent. Math.* **29** (1975): 203–204.

[17] R. Bowen and J. Franks, The periodic points of maps of the disk and the interval, *Topology* **15** (1976): 337–342.

[18] R. Bowen and D. Ruelle, The ergodic theory of Axiom A flows. *Inv. Math.* **29** (1975): 181–202.

[19] J. Brown, *Ergodic Theory and Topological Dynamics.* Academic Press, New York, 1976.

[20] J. Buescu and I. N. Stewart, Liapunov Stability and Adding Machines. *Ergod. Th. Dynam. Sys.* **15** (1995): 1–20.

[21] M. Caponeri and S. Ciliberto, Thermodynamic aspects of the transition to spatio-temporal chaos. *Physica D* **58**: 365–383, 1992.

[22] P. Chossat and M. Golubitsky, Symmetry-increasing bifurcation of chaotic attractors, *Physica D* **32**: 423–436, 1988.

[23] P. Collet and J.P. Eckmann, *Iterated maps of the interval as dynamical systems.* Birkhäuser, Basel 1980.

[24] C.C. Conley, Isolated invariant sets and the Morse index. *Conf. Board in Math. Sci.* **38**, 1978.

[25] M. Dellnitz, M. Golubitsky and I. Melbourne, The Structure of Symmetric Attractors. *Arch. Rational. Mech. Anal.* **123** (1993): 75—98.

[26] M. Dellnitz, M. Golubitsky and M. Nicol, Symmetry of attractors and the Karhunen-Loève decomposition. *Trends and perspectives in aplied mathematics*, 73–108. Appl. Math. Sci. **100**, Springer, New York, 1994.

[27] M. Denker, C. Grillenberger and K. Sygmund, *Ergodic Theory in Compact Spaces.* Lecture Notes in Mathematics vol. 527, Springer-Verlag, Berlin, 1976.

[28] J. P. Eckmann and D. Ruelle, Ergodic Theory of Chaos and Strange Attractors. *Rev. Mod. Phys.* **57**, 1985: 617–656.

[29] J.D. Farmer, Sensitive dependence on parameters in nonlinear dynamics. *Phys. Rev. Lett.* **55** (1985): 351–354.

[30] J. Franks and L.S. Young, A C^2 Kupka-Smale diffeomorphism of the disk with no sources or sinks. Proceedings of the Warwick Conf., Lecture Notes in Mathematics vol. 1007, Springer-Verlag, Berlin, 1980.

[31] L. Fuchs, *Infinite Abelian Groups* vol. 1, Academic Press, New York 1970.

[32] H. Furstenberg, Strict ergodicity and transformations of the torus, *Amer. J. Math.* **83** (1961): 573—601.

[33] J. M. Gambaudo, S. van Strien and C. Tresser, Hénon-like maps with strange attractors: there exist C^∞ Kupka-Smale diffeomorphisms on S^2 with neither sinks nor sources. *Nonlinearity* **2** (1989): 287–304.

[34] J.M. Gambaudo, D. Sullivan and C. Tresser, Infinite cascades of braids and smooth dynamical systems. *Topology* **33** (1994) 1: 84–94.

[35] J. M. Gambaudo and C. Tresser, Self-similar constructions in smooth dynamics: rigidity, smoothness and dimension. *Commun. Math. Phys.*, **150** (1992): 45–58.

[36] M. Golubitsky, I.N. Stewart, and D. Schaeffer, *Groups and Singularities in Bifurcation Theory*, vol. 2. App. Math. Sci. **69**, Springer-Verlag, New York, 1988.

[37] W.H. Gottschalk, Minimal sets: an introduction to topological dynamics. *Bull. Am. Math. Soc.* **64** (1958): 336–351.

[38] J. Guckenheimer, private communication.

[39] J. Guckenheimer, Sensitive dependence on initial conditions for one-dimensional maps. *Commun. Math. Phys.*, **70** (1979): 133–160.

[40] J. Guckenheimer and P. Holmes, *Nonlinear Oscillations, Dynamical Systems and Bifurcations of Vector Fields.* Springer-Verlag, New York 1983.

[41] J. Guckenheimer and P. Holmes, Structurally stable heteroclinic cycles. *Math. Proc. Camb. Phil. Soc.* **103** (1988): 189–192.

[42] J. Guckenheimer and R. F. Williams, Structural stability of Lorenz attractors. *Publ. Math. I.H.E.S.* **50** (1979): 59–72.

[43] P.W. Hammer, N. Platt, S.M. Hammel, J.F. Heagy and B.D. Lee, Experimental observation of on-off intermittency. *Phys. Rev. Lett.* (submitted), 1994.

[44] J.F. Heagy, N. Platt and S.M. Hammel, Characterization of on-off intermittency. *Phys. Rev. E* **49**, 2 (1994): 1140–1150.

[45] J.F. Heagy, T. Carroll and L. Pecora, Experimental and numerical evidence for riddled basins in coupled chaotic systems. *Phys. Rev. Lett.* **73**, 26 (1994): 3528–3531.

[46] M. Hirsch, Components of attractors. University of California at Berkeley, Preprint, 1992.

[47] M.W. Hirsch, *Differential Topology.* Springer-Verlag, Berlin 1976.

[48] J.G. Hocking and G.S. Young, *Topology.* Addison-Wesley, Reading MA 1961.

[49] F. Hofbauer and P. Raith, Topologically transitive subsets of piecewise monotonic maps which contain no periodic points, *Monatsh. Math.* **107** (1989): 217–239.

[50] M. C. Irwin, On the stable manifold theorem, *Bull. London Math. Soc.* **2** (1970): 196–198.

[51] M. Jakobson, Absolutely continuous invariant measures for one-parameter families of one-dimensional maps. *Comm. Math. Phys.* **81** (1981): 39–88.

[52] L. Jonker and D. Rand, Bifurcations in one dimension I: the nonwandering set. *Invent. Math.*, **62** (1981): 347–365.

[53] I. Kan, Open sets of diffeomorphisms having two attractors, each with an everywhere dense basin. *Bull. Am. Math. Soc.* **31**: 68–74, 1994.

[54] J.L. Kaplan and J.A. Yorke, Chaotic behaviour of multidimensional difference equations. In *Functional differential equations and approximations of fixed points*, (H.-O. Peitgen and H.-O. Walther, eds). Lecture Notes in Maths vol. 730, Springer-Verlag, 1979.

[55] Y. Katznelson, The action of diffeomorphisms of the circle on the Lebesgue measure. *J. Anal. Math.* **36** (1979): 156–166.

[56] G. King and I. Stewart, Symmetric Chaos. In *Nonlinear equations in the applied sciences* (W. F. Ames and C. Rogers, eds.): 257–315. Academic Press, London, 1992.

[57] L. Kocarev, A. Shang and L.O. Chua, Transitions in dynamical regimes by driving: a unified method of control and synchronisation of chaos. *Intl. J. Bifurcation and Chaos* **3** (2): 479–483, 1993.

[58] K. Kuratowski, *Introduction to Set theory and Topology.* Pergamon Press, Oxford, 1972.

[59] Y. Lai and C. Grebogi, Characterizating riddled fractal sets. *Phys. Rev. E* **53**, 2 (1996): 1371–1374.

[60] A. Lasota and J.A. Yorke, On the existence of invariant measures for piecewise monotonic transformations. *Trans. Am. Math. Soc.* **186** (1973): 481–488.

[61] F. Ledrappier and J.M. Strelcyn, A proof of the estimation from below in the Pesin entropy formula. *Erg. Th. Dynam. Sys.* **2** (1982): 203–219.

[62] A. M. Liapunov, The general problem of the stability of motion. Translated by A. T. Fuller. *Int. J. Control* **55** (1992), 3: 531–773.

[63] R. S. MacKay and J. D. Meiss, *Hamiltonian Dynamical Systems.* Adam Hilger, Bristol and Philadelphia 1987.

[64] R. Mañé, A proof of the C^1 stability conjecture. *Publ. Math. I.H.E.S.* **66** (1988): 161–210.

[65] R. Mañé, *Ergodic Theory and Differentiable Dynamics.* Springer-Verlag, Berlin 1987.

[66] I. Melbourne, An example of a non-asymptotically stable attractor. *Nonlinearity* **4** (1991): 835–844.

[67] I. Melbourne, Generalizations of a result on symmetry groups of attractors. *Pattern formation: symmetry methods and applications*, 281–295. Fields Inst. Commun. **5**, AMS, Providence, RI, 1996.

[68] W. de Melo and S. van Strien, *One dimensional dynamics.* Springer, 1994.

[69] J. Milnor, On the concept of attractor. *Commun. Math. Phys.* **99** (1985): 177–195.

[70] J. Milnor, On the concept of attractor: correction and remarks. *Commun. Math. Phys.* **102** (1985): 517–519.

[71] J. Milnor and W. Thurston, *On iterated maps of the interval.* Lecture Notes in Mathematics vol. 1342, Springer, Berlin - New York, 1988.

[72] M. Misiurewicz, Structure of mappings of an interval with zero entropy. *Publ. Math. I.H.E.S.* **53** (1981): 5–16.

[73] M. Misiurewicz, Absolutely continuous measures for certain maps of the interval. *Publ. Math. I.H.E.S.* **53** (1981): 17–51.

[74] L. Mora and M. Viana, Abundance of strange attractors. *Acta Math.* **171** (1993), 1: 1 – 71.

[75] S. Newhouse, Hyperbolic limit sets. *Trans. Am. Math. Soc.* **167** (1972): 125–150.

[76] S. Newhouse, Diffeomorphisms with infinitely many sinks. *Topology* **13** (1974): 9–18.

[77] S. Newhouse, The abundance of wild hyperbolic sets and nonsmooth stable sets for diffeomorphisms. *Publ. Math. I.H.E.S* **50** (1979): 101–151.

[78] S. Newhouse, Lectures on dynamical systems. In *Dynamical Systems*, CIME lectures, Bressanone. Birkhäuser, Basel 1980.

[79] H. Nusse and J. A. Yorke, Analysis of a procedure for finding numerical trajectories close to chaotic saddle hyperbolic sets. *Ergod. Th. Dynam. Sys.* **11** (1991): 189–208.

[80] V.I. Oseledec, A multiplicative ergodic theorem: Liapunov characteristic numbers for dynamical systems. *Trans. Mosc. Math. Soc.* **19** (1968): 197–231.

[81] E. Ott, J.C. Sommerer, J.C. Alexander, I. Kan and J.A. Yorke, Scaling behaviour of chaotic systems with riddled basins. *Phys. Rev. Lett.* **71** (1993): 4134–4137.

[82] E. Ott, J.C. Sommerer, J.C. Alexander, I. Kan and J.A. Yorke, A transition to chaotic attractors with riddled basins. *Physica D* **76** (1994): 384–410.

[83] E. Ott and J.C. Sommerer, Blowout bifurcations: the occurrence of riddled basins and on-off intermittency. *Phys. Lett. A* **188** (1994): 39–47.

[84] J. C. Oxtoby, *Measure and Category.* Springer-Verlag, New York, 1971.

[85] J. Palis and W. de Melo, *Geometric Theory of Dynamical Systems — an Introduction.* Springer-Verlag, New York, 1982.

[86] J. Palis and F. Takens, *Hyperbolicity and sensitive chaotic dynamics at homoclinic bifurcations.* Cambridge University Press, Cambridge, 1993.

[87] L.M. Pecora and T.L. Carroll, Synchronisation in chaotic systems. *Phys. Rev. Lett.* **64** (1990): 821–824.

[88] M. Peixoto, Structural stability on two-manifolds. *Topology* **1** (1962): 101–120.

[89] Y.B. Pesin, Characteristic Liapunov exponents and smooth ergodic theory. *Russian Math. Surv.* **32** (1978): 55–114.

[90] A.S. Pikovsky, On the interaction of strange attractors. *Z. Phys. B*, **55**: 149–154.

[91] A.S. Pikovsky and P. Grassberger, Symmetry breaking of coupled chaotic attractors. *J. Phys. A*, **24** (1991): 4587–4597.

[92] N. Platt, S.M. Hammel and J.F. Heagy, Effects of additive noise on on-off intermittency. *Phys. Rev. Lett.* **72**: 3498–3501, 1994.

[93] N. Platt, E.A. Spiegel and C. Tresser, On-off intermittency; a mechanism for bursting. *Phys. Rev. Lett.* **70** (1993): 279–282.

[94] H. Poincaré, Sur les courbes définies par les équations différentielles. *Jour. Math.* 3e série, vols. 7 (1881): 375–422, and 8 (1882): 251–296; 4e série, vols. 1 (1885): 167–244 and 2 (1886): 151–217. In *Oeuvres*, Gauthier-Villars, Paris, 1951.

[95] M. Pollicott, *Lectures on Ergodic Theory and Pesin Theory on compact manifolds.* Cambridge University Press, Cambridge, 1993.

[96] D.A. Rand, H. Wilson and J. McGlade, Dynamics and evolution: evolutionarily stable attractors, invasion exponents and phenotype dynamics. *Phil. Trans. R. Soc. Lond. B* **343** (1994): 261–283.

[97] C. Pugh and M. Shub, Ergodic attractors. *Trans. Am. Math. Soc.* **312** (1989): 1–54.

[98] H. J. Royden, *Real Analysis.* The MacMillan Co., New York, 1963.

[99] W. Rudin, *Real and Complex Analysis*, 3rd edition. McGraw-Hill Book Co., Singapore, 1986.

[100] D. Ruelle, A measure associated with Axiom A attractors. *Amer. J. Math.*, **98** (1976): 619–654.

[101] D. Ruelle, Analyticity properties of the characteristic exponents of random matrix products. *Advances in Math.* **32**: 68–80, 1979.

[102] D. Ruelle, Ergodic theory of differentiable dynamical systems. *Publ. Math. I.H.E.S.* **50**: 27–58, 1979.

[103] D. Ruelle, Small random perturbations and the definition of attractor. *Comm. Math. Phys.* **82** (1981): 137–151.

[104] D. Ruelle, *Chaotic evolution and strange attractors.* Lezione Lincee, Cambridge University Press, 1989.

[105] H.G. Schuster, S. Martin, and W. Martienssen, New method for determining the largest Liapunov exponent of simple nonlinear systems. *Phys. Rev. A* **33** (1986): 3547–3549.

[106] P.A. Schweitzer, Counterexamples to the Seifert conjecture and opening closed leaves of foliations. *Ann. Math.*, **100** (1974): 368–400.

[107] M. Shub, Endomorphisms of compact differentiable manifolds. *Amer. Jour. Math.* **91** (1969): 175–199.

[108] M. Shub, *Global Stability of Dynamical Systems.* Springer-Verlag, New York, 1987.

[109] M. Shub and D. Sullivan, Expanding endomorphisms of the circle revisited. *Erg. Th. Dynam. Sys.* **5** (1985): 285 – 289.

[110] G.F. Simmons, *Introduction to Topology and Modern Analysis.* McGraw-Hill, London, New York, 1963.

[111] S. Smale, Dynamical Systems and the topological conjugacy problem for diffeomorphisms. Proc. Int. Congr. Mathematicians, 490–496. Inst. Mittag-Leffler, Djursholm, 1962.

[112] S. Smale, Structurally stable systems are not dense. *Amer. J. Math.* **88** (1966): 491–496.

[113] S. Smale, Differentiable dynamical systems. *Bull. Amer. Math. Soc.*, **73** (1967): 747–817.

[114] D.R. Smart, *Fixed Point Theorems.* Cambridge University Press, Cambridge, 1974.

[115] M. Spivak, *A Comprehensive Introduction to Differential Geometry*, vol. I. Publish or Perish, Berkeley, 1970.

[116] K. Sygmund, Generic properties of invariant measures for Axiom A diffeomorphisms. *Inv. Math.* **11**: 99–109, 1970.

[117] S. Venkataramani, B. Hunt and E. Ott, Bubbling transition. *Phys. Rev. E* **54**, 2 (1996): 1346–1360.

[118] P. Walters, *An Introduction to Ergodic Theory*. Springer-Verlag, New York, 1982.

[119] J. Willms, Asymptotic behaviour of iterated piecewise monotonic maps. *Erg. Th. Dyn. Sys.* **8** (1988): 111–131.

[120] T. Yamada and H.Fujisaka, Stability theory of synchronised motion in coupled-oscillator systems II. *Prog. Theor. Phys.* **70** (1983): 1240–1248.

[121] L. S. Young, Bowen-Ruelle measures for certain piecewise hyperbolic maps. *Trans. Am. Math. Soc.* **287** (1985), 1: 41–48, 1985.

Index